How to Find a Higgs Boson–
and Other Big Mysteries in the World of
the Very Small

How to Find a Higgs Boson– and Other Big Mysteries in the World of the Very Small

IVO VAN VULPEN

Translated from the Dutch by David McKay
Illustrations by Serena Oggero

Yale UNIVERSITY PRESS

New Haven and London

Published with assistance from the foundation established in memory of James Wesley Cooper of the Class of 1865, Yale College.

Yale University Press books may be purchased in quantity for educational, business, or promotional use. For information, please e-mail sales.press@yale.edu (U.S. office) or sales@yaleup.co.uk (U.K. office).

Set in Minion type by Newgen North America, Austin, Texas.
Printed in the United States of America.

Library of Congress Control Number: 2019946667
ISBN 978-0-300-24418-2 (hardcover : alk. paper)

A catalogue record for this book is available from the British Library.

This paper meets the requirements of ANSI/NISO Z39.48-1992 (Permanence of Paper).

10 9 8 7 6 5 4 3 2 1

Contents

Introduction

"A universe can't be taken for granted. After all, there could just as well have been no universe." This strange, confusing statement sounds like the perfect opening to a profound philosophical reflection or a contest of wits at a cocktail party. But these words were part of physicist Paul de Jong's inaugural lecture when he became a full professor at the University of Amsterdam. In his lecture, de Jong confronted the ultimate question: *why is there anything at all?* This simple question is astonishingly deep, but if you think only about the *why*, you won't make much progress. Considering that something—the universe—does happen to exist, you might instead take a more pragmatic approach and focus on what that something is like, trying to figure out *how* everything works in this peculiar cosmos of ours.

Of course, you could lead a carefree life and live to be one hundred without ever asking yourself how electricity is produced, why water is transparent and stone is not, where the sun gets its energy, or how it is possible that the universe exists at all. Only after you've asked *why* for the first time, often by accident, do you notice how little you know about things that seemed so logical and trivial at first. The answer to most of those questions is only a mouse click away, but there are also many questions that remain unanswered. Nature doesn't just give away its deepest secrets without a fight. Generations of scientists like me have put all their effort into unraveling those secrets bit by bit. Through careful study and description of the phenomena we observe around us, we try to identify patterns that lead us, one step at a time, to the underlying laws of nature—because that is where the answers to all our questions lie hidden.

How does it all work? And why? To advance the frontiers of our knowledge, we must by definition enter territory where no one has gone before. The idea of going beyond the limits, in any field, has an irresistible pull. Sports records were made to be broken, extremes of nature are just asking to be conquered, and we remember the names of the first adventurous people to reach great milestones. In 1968, Jim Hines was the first athlete to run the hundred-meter dash in less than ten seconds; Edmund Hillary and Tenzing Norgay arrived at the summit of Mount Everest in 1953; and Jacques Piccard explored the depths of the ocean in 1960, descending more than ten kilometers into the Mariana Trench. In 1911, Roald Amundsen was the first to reach the South Pole; in 1969, Neil Armstrong did the same on the surface of the moon. These were all immortal heroes who gave imagination a face by advancing our frontiers. They expanded our world, both literally and figuratively, and we all travel in their footsteps. On to the next challenge!

These pioneers wanted to be the best, the first, or the fastest, and scientists are no different. Scientists, too, look longingly at the horizon and set goals that seem unattainable to others. They too are driven by the desire to be the first to answer the questions that have preoccupied so many before them. They stand in a long line of scholars who, since time immemorial, have pursued the ultimate, elusive "Why?"

While we associate adventure and progress with words like "bigger," "better," and "farther," some scientists are on a quest for the opposite extreme: small, smaller, smallest. We all know that children can make the most amazing structures out of Legos, from a simple tower or castle to a huge spaceship or an entire fantasy world. Yet whatever they make, it is composed of a small set of distinct building blocks. We find exactly the same thing in nature. The whole wondrous array of highly complex objects we see in the world around us, from a star to a human being, from a drop of water to a virus, is composed of the same small set of building blocks. The scientists who search for those elementary particles of nature are called particle physicists. Their goal is to understand how these particles combine to form our everyday world in all its intricacy. This too is an adventure—a breathtaking journey of discovery, like descending a spiral staircase that winds deeper and deeper into the foundations of the universe.

This book describes the tireless search of the many scientists who for hundreds of years have been tracking down the ultimate building blocks of nature and everything they hold within. That search sometimes resembles an exhausting, endless cycle of questions, answers, and new questions. We have certainly come closer to the ultimate answer. Yet as impressive as our present understanding of nature may be, one thing is certain: the universe and its laws are completely crazy.

For the past hundred years, we have slowly been making our way down that spiral staircase, like archaeologists unearthing ever deeper structures. And each new level we discover is like a new country, with a wealth of knowledge and insights that offer us revolutionary perspectives on nature and answers to age-old questions. But our new discoveries also raise new questions, exposing mysteries and dreams from still deeper levels of reality. It is clear that we have not yet reached the bottom.

In this book, I will take you as far as I can down the staircase, into the world of particle physics, the world to which my work is devoted. But we will have to proceed with caution. It is a world invisible to human eyes, dominated by seemingly magical behaviors and scientific instruments like quantum mechanics, the theory of relativity, and particle accelerators. These terms may sound intimidating to a non-physicist. But don't despair. I have written this book for anyone who's ever wondered about the discoveries described in the newspaper. As we descend the staircase of nature's building blocks, I'll not only tell you about the discoveries we've made along the way, but also show you how we managed to take each new step and how those discoveries have changed society.

Even though no paleontologist has ever seen a dinosaur, or ever will see one, examining bones and other remains has taught us an enormous amount about a world that vanished nearly a hundred million years ago. Particle physicists are in a similar position. The world we explore seems as far beyond human reach as the dinosaur world is to paleontologists. That's because we're interested in a world millions of times too small to see with the naked eye. But by carefully fitting together the remains of collisions between particles, as if we were assembling loose dinosaur bones into complete skeletons, we can penetrate ever deeper into that minuscule world.

This untiring journey to the foundations of our universe has also pushed technology to its limits and beyond, as scientists create new ways to consider these questions. And although few people truly grasp the mathematical formulations and deeper implications of the new insights, the inventions and technologies developed for this quest have become omnipresent in our lives. In fact, they are already part of the fabric of our modern society. Without the theory of relativity there could be no GPS; without quantum mechanics, no computer chips; without antimatter, no PET scans to locate tumors; and without particle accelerators, no way of irradiating the malignant tumors we find.

This branch of science is far removed from the traditional image of a solitary old man in a dusty laboratory. It has become a global enterprise, in which scientists from every country have no choice but to work together across political divides at large research institutes. One such organization—CERN in Geneva, the European Laboratory for Particle Physics—is the setting for much of this story. It is not only a research institute, but also a large-scale sociological experiment, since it's not easy to persuade even a small team of stubborn physicists to collaborate, let alone thousands of physicists from more than one hundred countries. But thanks to the dream they share— a dream of learning more about the fundamental building blocks, and the past and future of the universe—these peculiar physicists have succeeded, against all odds, in advancing the frontiers of our knowledge and unearthing one new secret after another.

Particle physics is an adventure driven by a hunger for answers to big questions about the workings of our universe. This book will show you how amazingly far we have already come. Although the deepest layer of reality reveals that the matter in our universe is composed of just a few distinct building blocks, it also confronts us with patterns and phenomena that utterly bewilder us. In 2012, for instance, we discovered that empty space, the stage on which the planets and stars

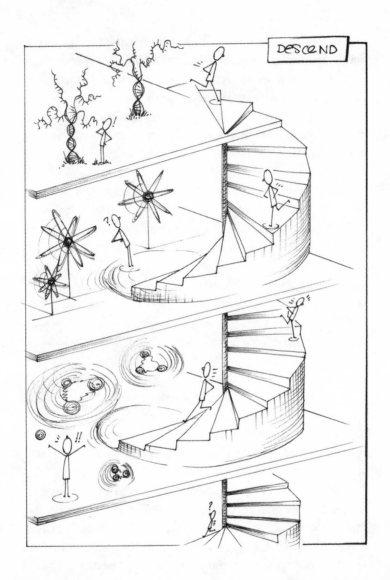

appear, is not really empty at all, but filled with a mysterious field. A field that gives all the building blocks mass, which makes them clump together into objects like the sun and the earth. Yet so much remains unclear. What mechanism is responsible for the mysterious yet unmistakable patterns we find in the world of elementary particles? And what about the origins of space itself? We're still stumped. There must be a hidden door somewhere, leading down to a still deeper foundation, a more profound truth that answers all those open questions. But where is that door, and how can we step through it? Off we go to find out!

1

The Rules of the Game

We begin our journey of discovery on the ground floor, in the familiar world of things we can see and hold in our hands. A world where we know our way around and can make careful preparations for our descent. For example, we'll need to understand how scientists discover physical laws by studying patterns in nature. It's also important to see that when we search for answers to the big questions, the laws we discover, however simple, can have a huge impact on our society. Sometimes the patterns are perfectly clear but the answers are hidden at a level invisible to the naked eye. How on earth can you see something a million times smaller than the smallest thing visible to the naked eye (spoiler: with a particle accelerator) and how do you learn your way around that strange world? We'll begin our adventure on solid ground before we start digging.

The Starting Point

Over the past centuries, we have uncovered many of nature's secrets. But how do you actually do that? An apple can't tell you why it falls when you let go, and you can't ask the sky why it's blue. Instead, to learn how nature works and unearth its secrets, you have to study the universe very systematically. How does nature behave under various conditions, ordinary or extreme? What phenomena do we observe, in all their details?

Instead of just sitting around waiting for nature to offer you information, you can roll up your sleeves and create a variety of conditions in a controlled way. That's what experiments are: a way of asking nature to answer the questions we have about its behavior. Experiments produce a wealth of facts that we can document and categorize. And they often show us that however complex the phenomena of nature may seem at first, they can usually be traced back to a small number of deceptively simple underlying principles. To find your way to these deeper rules and laws that nature strictly obeys, you have to be able to recognize patterns in the facts that you've collected. Children use this strategy almost instinctively when trying to understand the world around them. How will my parents react if I draw on the wall with my felt-tip pen? What will happen if I suddenly scream while we're waiting in line at the supermarket? And will it really hurt if I stick my finger in the flame?

Although parents may react in all sorts of ways (as I can tell you from experience), nature responds according to unbreakable, ironclad principles. These regularities, the laws of nature, give us information about how things work, and they're universally applicable. As soon as you've figured out how nature behaves, you can predict how it will behave in the future and in other situations. That's how, over the years, we've gradually learned more about the world.

Making scientific progress is not as easy as it may sometimes seem in retrospect. By definition, scientists are almost always in unknown territory, driven by an irresistible urge to find answers and not even knowing for certain that answers exist. Real scientific progress comes in spurts. We usually move forward in small steps, but every now and then comes one of those rare moments when we suddenly make a great leap. At a moment like that, we recognize the underlying mechanism and hit on a more fundamental set of laws. After the initial euphoria, we take our first careful steps into this new reality, this new world. And when we do, we find time after time that we observe new phenomena there. This kind of breakthrough could be the result of some genius's new insight, but it could just as easily be a chance discovery, or the outcome of a new experimental technique that allows us to study nature at a very different level. One good example is the invention of the microscope, a new technique that revealed a host of living mechanisms concealed in something as simple as a drop of water and so brought to light a hidden world.

This invention was a crucial step forward for medical science. And this type of discovery, which unearths a deeper level, takes us from the *how* to the *why*, as we learn to understand the strange phenomena we previously observed in terms of the new laws we've just discovered. But that's not all it does for us. It also often gives us the tools to make connections between phenomena that up to that moment had seemed completely unrelated.

Before we embark on our journey of discovery into the abstract world of elementary particles, I'd like to give a few examples of how we've persuaded nature to spill its secrets. Sometimes it was easy, sometimes hard, but each example shows that the drive to understand phenomena has both irreversibly transformed our understanding of nature and produced the knowledge on which our modern society is based. For the rest of this book, our guiding theme will be the gradual discovery of the world of elementary particles. But each time we look at an insight or discovery, however abstract and fundamental it may be, I will also try to show its practical applications, which have become indispensable parts of our everyday lives. We will see that fundamental research not only brings deeper insight into the workings of nature, but also has a profound long-term influence on the economy and society.

Going on a journey of discovery means being the first to enter new terrain, so you're almost certain to run into unexpected problems that don't yet have a solution. For example, no matter how well you can build strong, sturdy bridges, if you want to know what's on the far side of the ocean, you really have no choice but to build a boat. Other times, it's not hard work that leads to the next step, but a clever idea. To find out what's inside a walled-off area, you could chip away at the wall with a hammer and chisel for years, but the smart thing to do is make a ladder. This all seems deceptively simple, because we already know how to solve these particular problems, but imagine being the first person ever to come up with these ideas.

In short, scientists are real adventurers. They may not become millionaires, but imagine the triumph and everlasting fame of being the first to reach the summit of Mount Everest or set foot on the moon, or, in my own field, of being the first to identify the fundamental building blocks of nature, figure out why no antimatter can be found anywhere in the universe, or learn whether empty space is truly empty or actually full of a mysterious substance that gives all particles their mass.

There's a great analogy I once heard from a colleague that sheds light on how challenging it can be for a scientist to identify patterns and construct a theory. Imagine you're an alien who has just landed on earth. Chances are almost everything you see on this planet surprises you, but you decide to be systematic and start with something simple. So you ask yourself, "What are the rules of the popular game soccer (or "football") played in every country on this planet?" It's a clear question and may seem easy to answer. But there's a catch: you can watch as many games as you like, but you can't talk to anyone about it or read anything on the subject. All you can do is watch. Give that a minute to sink in, and then ask yourself how long it would take for you to come up with a complete list of the rules.

You'd probably figure out pretty fast that there are two teams of eleven players, that the whole game is played between the outermost white lines, that players switch sides after forty-five minutes, and that the basic objective of the game is to kick as many balls as possible into the opponent's goal. But who are those players, one on each side, who do a lot less running, dress differently from their teammates, and are allowed to use their hands? How long will it take you to realize what the two humans are doing who run back and forth along the sidelines with little flags, or to understand corner kicks, offside, substitutions, the mysterious extra time at the end of some matches, the strange lines on the playing field, penalty kicks, and so on? Imagine how difficult it would be to discover all the rules. Difficult, but not impossible—at least if you're really, really motivated and willing to invest enormous amounts of time. Scientists face exactly the same challenge, but this time, the playing field is the world around us. Nature doesn't just give away its secrets free of charge. Only by observing carefully, and designing experiments that ask nature the right questions, can we figure out what phenomena exist. That allows us to decipher nature's rulebook, bit by bit.

By the way, no one ever said the rules had to be logical. In fact, the laws of nature *don't* fit everyday logic—not one of them. Quantum mechanics and relativity, two of the most famous theories we'll encounter in this book, are both deeply weird. In a sense, you could compare them to the offside rule in football: absurd but true, and simply the way the game is played. As soon as you accept that it's a rule, it's only logical that certain goals count and others don't. Likewise, the bizarre principles that underlie theories like relativity and quantum mechanics are completely counterintuitive, but once you accept them, they do explain all the bizarre and complex phenomena that we see when we look at nature on the atomic scale. The theories are right. But logical? Nope.

By working out the consequences of these strange theories, we've learned how to apply them in useful ways in everyday life. Much of the current research and progress in nanotechnology and quantum computing are founded entirely on that peculiar theory known as quantum mechanics. Although many experimental results and observed phenomena are "logical" according to that theory (in other words, they're explained by the strange laws of quantum mechanics), there's not a scientist on earth who understands *why* the world obeys the quantum mechanical laws. For instance, how is it possible for a particle to be in two places at once or in an entangled state that would be unthinkable in our ordinary world? You can't be both pregnant and not pregnant at the same time, but in the quantum world, mixed states of that kind are completely normal.

The more successful a theory is, the more comfortable we get with the rules and laws of that strange world. At the same time, it's frustrating to have no explanation for the basic building blocks of your logical framework. In this case, we've gone beyond the *how* to the *why*: quantum mechanics. But no sooner do we take that step than the question shifts from the *how* of quantum mechanics to the *why* of quantum mechanics. In other words, the answer immediately gives rise to a new question. The poor scientists are always running after shifting goalposts. But their insatiable curiosity has led to huge gains for our society, because the insights and applications arising from their work have become foundation stones of our civilization. Still, it's important to realize that even the smartest scientists on earth will run out of explanations fairly quickly if you keep asking "why."

Over the past hundred years or so, elementary particle physicists have managed, step by step, to reach the innermost depths of the atomic nucleus. In the course of that adventure, we have taken a few great strides and arrived at fascinating insights—not only into the building blocks of all the matter in the universe, the elementary particles, but also into the fundamental forces of nature. There's every reason to take pride in that. Before we delve into the world of elementary particles, I'd like to show you a few simple laws of nature that we humans have learned to manipulate, and patterns that we've uncovered. These three examples will demonstrate that a few things we take for granted have no logical basis but nevertheless play a crucial role in everyday applications that arose from pure fundamental research. The first two relate to the questions of how electricity is generated and where inherited traits are encoded in the human body.

One of the greatest threats to our prosperity and way of life is a shortage of energy. We don't often pause to think about it, but our Western society is addicted to energy, and without electricity, it would come to a complete standstill in less than a day. Try to imagine a typical working day without electricity: no alarm clock, no lamps, no coffee machine, no car, no elevator, no ATMs, no television, no computer, no radio, no Internet, no telephone, and no dishwasher. You can clearly see why energy is such a hot topic of public and political debate. This book is not the place to review all the facets of this complex issue, such as the earth's finite supply of fossil fuels, the geopolitical interests at stake, carbon dioxide emissions, green energy, and the nuclear power debate. You could fill a library with expert studies of those subjects, and the debates are still in full swing. My job here, as a physicist, is to raise a question that I know doesn't play a central role in the discussion, but which I'd like everyone to be able to answer: *How do you make electricity?* For example, how do you turn a charcoal briquette, the kind you use in your barbecue grill, into an electrical current? We can do that thanks to a secret that nature revealed to us through one simple observation. The discovery of this seemingly straightforward law of nature changed our lives and our civilization fundamentally.

Some one hundred and fifty years ago, James Maxwell managed to capture all known facts about electricity and magnetism in four famous formulas that have carried his name ever since: the Maxwell equations. They show that magnetism

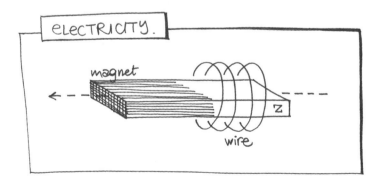

and electricity are intimately interrelated, and they describe electromagnetic phenomena that can be extremely complex—including a regularity discovered earlier by Michael Faraday that gives us a way of making electricity.

Observation (and law of nature): *A current starts running through a coil of copper wire when the magnetic field in the center changes.*

That may not sound so exciting, but if you take a coil of copper wire and move a magnet through it, it really happens. The magnetic field is absent at first, becomes very strong as you move the magnet into the center of the coil, and then disappears once the magnet has been removed completely. This change creates an electrical current in the wire. That's the simplest way I can put it. But I couldn't make it much more complicated either. So there it is: the principle that allows us to generate electricity, whether in a bicycle lamp or in the most up-to-date nuclear power plant.

When you ride a bicycle, you generate the power for your own lamp. A hub dynamo is basically a very long, rolled-up copper wire, like the long iron wire coiled neatly around the hose of a vacuum cleaner. Inside that tube of coiled wire is a magnet connected to your wheel by a gear called a roller. As you pedal your wheel turns, and so does the magnet. And because of the law of nature we discovered, which says that changing the magnetic field in the tube of copper wire will make a current run through it, a current really does run through it. This current is then conducted through a thin metal wire into the lamp, making it warm up and start to glow. Ta-da: a bicycle light. And not to make the high-tech energy giants seem unimpressive, but a big coal-fired power plant works about the same way. There too, a magnet is rotated inside a coil of copper wire. The only difference is how you move the magnet. On your bicycle you do that by pedaling, while in a power plant the work is done, surprisingly enough, by a turbine. The turbine spins because of steam pushing hard against the blades, which are connected to the magnet by a gearbox. So how do we make steam? By heating up a large container of water. And how do we heat the water? By burning a heap of coal underneath it. It's that simple.

Of course, thousands of people work hard every day to make each step of this process as efficient as possible in power plants, and much more is involved than I describe here, but this is the basic principle. And a nuclear power plant works almost the same way. The only difference is how you heat the water. In a nuclear plant, this is done by the particles released when you split the nuclei of heavy atoms such as uranium. Wind energy uses the same principle: the spinning blades of the turbine turn a magnet inside a copper coil.

This very simple principle—generating current by changing a magnetic field—is fundamental to our economy and therefore to our prosperity. When Faraday first made these discoveries, no one could suspect how they would be applied. William Gladstone, who was in charge of the British Treasury, is said to have asked Faraday, "But, after all, what use is it?"— an understandable question from his point of view. It is the same question that scientists are still asked today whenever we apply for research funding. Unfortunately, we can no longer get away with Faraday's famous reply: "Why, sir, there is every probability that you will soon be able to tax it!" Back then, people were fairly content with candlelight, and it probably seemed more sensible to give money to the candle industry to find a more efficient production method or design a better wick. But in retrospect, we can see that we would never have discovered the light bulb that way.

Even though a lot of scientific research leads absolutely nowhere, this is a striking example of how true innovation can't always be planned in advance. The game-changing discoveries often come from unexpected places. That's an important message for politicians and for society as a whole: alongside innovation for industry, we need to create enough opportunities for free, unrestricted fundamental research. Applications are sure to follow.

For physicists, it's often both intriguing and frustrating to follow public debates about electric cars and hydrogen-fueled vehicles. We're sometimes startled to realize that, despite all the well-meant efforts of scientists and manufacturers, politicians and policymakers are unaware of the most basic scientific facts. The discussion often focuses on distant future scenarios, but a little more technical savvy among politicians could have a greater impact than all the scientific reports in the world. True, a Tesla doesn't give off carbon dioxide as you drive, but even so, it's striking how rarely people think to ask where the energy in the battery comes from. That battery is charged, via an electrical socket, with power generated by a coal-fired station that produced lots of carbon dioxide. On top of that, the battery is made of heavy metals and very aggressive acids, not exactly "green" technology. Of course it's a good idea to reduce our consumption of fossil fuels, and of course a large power plant is more efficient than a thousand separate automobile engines. And of course you can also charge the Tesla battery with solar energy. Even so, the popular belief that electric cars are squeaky clean is an exaggeration.

A similar debate is in progress about the hydrogen economy. The idea is to generate power by mixing hydrogen and oxygen, stored in separate tanks, and then burning them as fuel. That's certainly an ultra-clean process; it produces nothing but energy and water. You might think it's the ultimate in clean fuel for your car, but again, there's a catch. Where do you think you find the pure hydrogen and pure oxygen? Well, you start with water, which is made up of oxygen and hydrogen atoms. To separate the two, you need energy. It's exactly the same process as combining them for fuel, but in the opposite direction. And where do you find the energy for doing that? That's right—usually a large coal-fired power station or a nuclear plant.

Or from a wind turbine, of course. There's no denying that we can also use green energy to separate hydrogen and oxygen from water. But my main point here is that hydrogen is an energy *carrier*, rather than an energy *source*. It does offer certain advantages: no carbon dioxide or soot is released in the area where the car or bus is driven, the city center, and it's a fantastic way of storing excess energy from turbines, solar cells, and power plants for short-term use. But it's not *the* solution to the energy problem.

Not only physicists, but scientists in many fields, are in search of explanations for undeniable patterns that they see but don't understand. Fundamental research has not only given us electricity but also led to major discoveries that are now central to medical science. When you look at nature systematically, you always find a treasure trove of data. Sometimes, once you've collected enough information, you can recognize patterns and find your way to deeper insights. The technique of reductionism—zooming in on the basic building blocks of which things are made—is not reserved exclusively for physicists. One good example of the progress we've made by probing ever deeper into the building blocks of cells is the discovery of DNA and the coding of genetic information. This discovery stands head and shoulders above all others in its impact on medical science.

Biologists and farmers have known for a very long time that animals and other organisms pass on traits to their offspring. The best-known example is probably the eye color of parents and children. If a child's parents both have brown eyes, the chance that the child will have brown, green, or blue eyes is 75 percent, 19 percent, or 6 percent, respectively. There are many tables of this kind for different traits, from the color of a cat's fur, to the resistance of crops to certain diseases, to the ability of plants to adapt to saline soil or other extreme conditions. In the case of eye color, the only serious problem arises when we need to explain why a child has brown eyes even though both parents have blue eyes (since the probability of that is zero). But heritable traits do have a tremendous influence on our food supply. In the agricultural sector and the food industry, knowledge about heritable traits is used daily in an effort to pass on desirable traits, such as resistance to disease, high milk productivity for cows, or the adaptability of rice plants to extremely dry or wet climates. Selective breeding of plants and animals over many generations can make desirable traits more widespread in the population, using the laws of nature for our benefit.

So we can *see* that nature follows certain patterns, and we could devote our entire lives to studying which traits are and aren't heritable. But the question we'd most like to answer, of course, is: *How* does it all work? The source of individual traits is apparently hidden away somewhere in the body. But *where*? Is it only in the egg cell and sperm, or in every one of the body's cells?

This question went unanswered until the 1960s, when new detection techniques enabled Francis Crick, James Watson, and Rosalind Franklin (the last of whom is often "forgotten" in the history books) to investigate structures much smaller than a human cell. They discovered the double helix structure in which genetic information is stored: DNA. Information about eye color and many other traits turned out to be kept in cell nuclei, and the language in which that information was written had an alphabet with only four letters, each corresponding to one nucleotide: C (cytosine), G (guanine), T (thymine), and A (adenine). Together, these four organic molecules encode all the traits and complex phenomena we observe in living things. What we learned through this revolutionary discovery is that, while our alphabet has twenty-six letters, only four are needed to record an individual's complete genetic code. If you were fluent in that language, you would know right away where a person's eye color is encoded, and you'd understand why some people are susceptible to certain diseases and others are not.

This was one of the most important scientific discoveries of all time and laid the groundwork for modern biomedical research and drug design. It yielded crucial insights into cell division and pointed to new questions: Where is each specific trait encoded? How does the cell copy the DNA when it divides? What is the effect of an "error" in a strand of DNA? How do you "read" the strand? What words are formed by the CTGA combinations? Can we find the source of cancer in the genes? And can we manipulate the genome to prevent disease? Although the basic idea of the alphabet of genetic material has been around for fifty years, we have not yet completely mastered the language. We find new combinations and patterns just about every week and only recently have succeeded in fully mapping large portions of the human genome.

But our knowledge is evolving at a dizzying pace. On the website of the U.S. National Human Genome Research Institute, you can find statistics showing that it cost a hundred million dollars to decode the entire genome in 2001. Today, it costs only a few thousand. And home kits can even be used to analyze part of your DNA from a sample of your saliva. Apart from medical scientists, physicists such as Cees Dekker in Delft, the Netherlands, are also participating in cutting-edge genetics research, and their work has led to the ability to read out long DNA strands efficiently. Once we can do that, the obvious step will be to start building our own DNA structures.

Like anything else, this development has a good side and a bad side. In recent years, advances in genetics have made the news almost every week. Sometimes it's a new genome that's been partly or fully decoded, or a genetic modification developed to track down or even repair the source of a disease. Everyone is in favor of new techniques for the early diagnosis of heritable disease, or new medicines tailored to individual genetics, but at the same time, these innovations give rise to many ethical debates. For example, do I want my health insurer to know that I have a high risk of cancer, and if so, what can we—or should we—do with that information? How will society handle the capability to detect many genetic disorders in unborn children, and do I have the right to decide for myself whether or not I want that information? While selective breeding of plants and animals for desired traits is pretty widely accepted, that's not the case with direct manipulation or synthesis of genetic material—in other words, genetic modification. The moral of this example is that simply compiling larger and larger books of tables for heritable traits would never by itself have led us to discover DNA; figuring out the pattern, the genetic alphabet, made the difference. Whether and how we use our new insights and techniques are matters of ongoing social debate.

Success stories about recognizing patterns, like the story of DNA, are an easy way for scientists to score points. But just as often, science is a frustrating tale of our inability to really understand the phenomena we study. Sometimes what we need is a brilliant insight, and sometimes we simply don't yet have enough information or knowledge to take the next step.

Unfortunately, you can't always find satisfactory answers to the questions you ask yourself about nature. If there's one thing we can count on, it's that the sun will rise again tomorrow. Right? It sounds so self-evident. But when Albert Einstein was young, he probably looked at the sun now and then and thought, "What keeps it burning?"

Now, I don't mean to pretend that I know what went on in Albert Einstein's head, but I know for certain he couldn't answer that question back then. What makes me so sure? Well, *no one* in the world knew the answer a century ago, because the scientific knowledge required to understand the answer, even in the most basic terms, was simply not yet available. Oddly enough, I never thought of that myself until about ten years ago, when I was preparing a talk for schoolchildren. It's a strange idea, because if no one knew what kept the sun burning, that means no one knew how long it had been burning and—not exactly an insignificant detail—how much longer it would go on burning. So what did Albert and his scientist friends believe? And what did the rest of humanity think? Why wasn't everyone on earth obsessed with the question?

Yes, of course, *now* we know that the sun keeps burning because energy is released whenever two hydrogen nuclei fuse into a helium nucleus in its core, thanks to the incredibly high temperatures there. But in the early twentieth century, the atomic nucleus had not yet been discovered. That didn't happen until thirty years later, when scientists pushed their machines to the outer limits of their capacity. That discovery, again, would inspire a variety of applications, as we'll see in

Chapter 2: not only nuclear energy and the atomic bomb, but also nuclear fusion, our great hope for a clean solution to the earth's energy problem.

Venturing into the World of the Very Small

Any expedition into unfamiliar territory demands the right equipment. If you want to go to the North Pole, you'll be better off with warm clothes, a pocketknife, and a dog sled than with a Hugo Boss suit, a cheese slicer, and a bicycle. And if you want to go to the moon, you'll need to build a rocket and find a spacesuit. Our branch of science is no different. To descend into the world of elementary particles and explore structures even smaller than the building blocks of DNA, we need one essential tool: the particle accelerator. This workhorse of particle physics is a complex piece of equipment that, like a Swiss pocketknife, can be used in various ways.

First of all, a particle accelerator is an extremely good microscope, with a power of magnification thousands of times greater than ordinary microscopes can ever achieve. Long before we had this tool, it became clear that there was a fundamental limit on the smallest objects we can see with a conventional microscope. But the particle accelerator finally allowed us to break through that seemingly impenetrable barrier into a much smaller world. We not only became familiar with the smallest distinct building blocks of each element, atoms, but also learned exactly how they were made up of even smaller building blocks. We also found out that on that scale, the laws of nature are fundamentally different than in our everyday world.

Encouraged by the success of this new technology, we built ever more powerful particle accelerators and discovered a second way of using them: as "nutcrackers." You see, a microscope is not always the right tool. If you want to know what's inside a walnut, for instance, then a microscope won't help. The microscope may show you the nutshell in incredible detail, but if you want to know what lies hidden within the shell, you'll have to break it open with a hammer or a nutcracker. And that's a perfect description of our second way of using a particle accelerator: by firing particles at an object at high velocity, we can actually crack open either those particles or their target. By studying the wreckage left by the collision, we can find out what was inside the material that we bombarded.

What will be most relevant to our journey in this book is the *third* way of using a particle accelerator: to create new matter. We've discovered—to our great surprise, I might add—that if you fire particles with extremely high energy at each other, they not only collide but also actually create *new* particles. This led to utter chaos at first, because we discovered hundreds of different kinds of tiny particles, but eventually the whole menagerie turned out to be a beautiful puzzle that could be assembled with a very limited set of building blocks. Those building blocks, and the ways they attract and repel each other, ultimately came together to form what we call the Standard Model: the outer limit of our knowledge of elementary particles, a magnificent framework that still stands tall and describes almost every phenomenon in that mini-world. We'll discuss this third function later on. But first, let's consider the particle accelerator in its role as a super strong microscope.

In everyday life, we use our eyes, nose, ears, mouth, and hands to perceive the world around us. Our eyes and nose, for example, are perfect instruments for distinguishing jam from butter in the morning. At the same time, we know that our eyes, as complex they are, fail us completely when we try to look at very small objects. Threading a needle is difficult enough;

forget about checking whether an ant has teeth or whether there are bacteria in a drop of water.

Long ago, we figured out how to combine lenses in an ingenious way to make a microscope, with which we can explore a smaller world. But even today's microscopes, which are far stronger than Antonie van Leeuwenhoek's original model, run up against a fundamental limit to their power. A microscope can never—*ever*—detect things smaller than a millionth of a meter. That's incredibly small, of course, and microscopes are perfect for looking at bacteria and cells, but they're just not the right tool for studying DNA or atoms. To break through that fundamental barrier, you need a clever trick. And we found one! Strangely enough, it required us to start looking at things without using our eyes.

Whenever we *see* something, it's because our eye has captured particles of light that have bounced off the object we're looking at. As you walk down the street, you can see the people around you because light from the sun bounces off them and straight into your eye. The retina in the back of your eye acts like a kind of digital camera with a whole lot of pixels, and your brain has learned to interpret the patterns and translate them into complex objects. That's how you can see the difference, at a glance, between a lamppost and a human being, or between stones and water. Our own built-in digital camera has two components: "rods" for measuring the light *level* and "cones" for seeing different *colors*. Together, these two types of photoreceptors in the human retina give us enough information to perceive the world around us.

But however sharp our eyes and ears may be, they're not perfect, and many worlds remain hidden from us. For example, there are pitches that our ears cannot detect but that dogs can hear perfectly. And besides good hearing, dogs have a phenomenal sense of smell. That's why they're used in harbors and airports to track down drugs and money hidden in suitcases. There are also worlds hidden from our vision—things that people can't see but that really are there. Our eye is not infinitely sensitive and can't see all colors.

We're all aware of our limits when it comes to the *intensity* of light. On a dark night, we humans can hardly see a thing, but we know that cats don't have that problem. Since cats have many more rods than cones, they can see even in very poor light. Their eyes work better because of their different structure. Cats see a whole world in the dark, a world hidden from humans because we simply aren't equipped to perceive it. But sometimes people do need to see in the dark—so, being the resourceful creatures that we are, we invented night vision devices. They intensify the few particles of light that do manage to reach us into a signal that we *can* see. Pretty smart!

Even when the sun is at its height, some worlds remain hidden from us simply because our eyes can't see every *color*. I'm not talking about color blindness, but about the limitations of the average human eye, which has cones that are sensitive only to the colors of the rainbow, the familiar range from red to violet. You see, each color of light corresponds to a distinct wavelength—meaning the length of the wave of light that flies toward you. The shortest waves that we see look violet, and the longest ones look red. Because of the shape of the rods in the human eye, they aren't at all sensitive to the colors corresponding to longer waves than red or shorter ones than blue. But those colors do exist. They're called infrared and ultraviolet, and plenty of animals can see them: an entire world hidden from us in bright daylight because our human eyes can simply not detect them.

A bee, for instance, can see ultraviolet light, those colors with a slightly shorter wavelength than violet, which is just barely in our visible range. Is that a useful skill? You bet. Along with the yellows and reds visible to us, some flowers display intense ultraviolet colors, which we can't see at all. So a bee flying over a field of grass can pick out the different flowers there effortlessly, while we struggle to make out a few vague hints of flowers in the green grass. At the opposite end of the spectrum are animals that can make out infrared colors, which we can't see with our eyes but can feel as warmth on our skin. Snakes use those colors to track down prey with ease.

We humans can invent special devices, like those clever night vision goggles, to use in scientific experiments. You might call them artificial eyes, ears, and noses that enable us to discover, observe, and explore worlds hidden from our senses. We must always remain aware that there's more around us than we can perceive with our senses and keep working hard on smart technology that makes those worlds visible to us in other ways.

A little more than four hundred years ago, a Dutch inventor of scientific instruments, Hans Lippershey, noticed that an ingenious combination of lenses made it possible to magnify a small object. The possible applications were endless. Galileo Galilei improved the telescope that allowed him to study the moon and the motions of the planets in great detail. But we also learned how to peer "into the depths" with microscopes, gaining access to another completely unknown realm. For example, Antonie van Leeuwenhoek discovered a world of wonders in something as simple as a drop of blood, becoming the father of modern microbiology. Although in the centuries since then we've steadily improved the design of the microscope, we knew that this avenue of exploration would one day come to an end, because of a fundamental limit to what you can see with light. Even using a microscope, we will *never* be able to see anything smaller than a millionth of a meter (about one hundredth of the thickness of a hair). That's not because we can't make better lenses, but because waves of light simply don't bounce off such small objects.

One thing we know about waves is that they bounce off only those objects larger than their own wavelength—a principle of physics that, for the time being, you'll just have to accept. Think of a marble rolling across the kitchen floor—it bounces when it hits the trash can, but rolls over a breadcrumb without any effect. To calculate the smallest object that light can bounce off, we need to know how large a light wave actually is. That varies a little by color—as I mentioned, red has a longer wavelength and blue a shorter one—but the light visible to us humans normally has a wavelength of a little bit less than a millionth of a meter. If you want to see something even smaller than that, you'll never succeed with a traditional microscope, not even the most powerful one in the world.

Fortunately, when faced with an insurmountable obstacle like this one, we don't have to throw in the towel. We just need a bright idea or a completely new approach. That's how fundamental science works: amazingly, no matter how big the problem is, a solution is eventually discovered that breaks through the barrier and expands our horizons. This problem was no exception. But we did have to look very far beyond traditional techniques.

Besides looking with your eyes, there are plenty of other ways to figure out an object's shape. If you close your eyes, you can still easily feel the difference between a knife and a fork.

Scientists use a similar method to "feel out" objects, but instead of using our hands, we fire little bullets and watch how those bullets bounce off the object we're studying. The way they rebound, or *scatter*, gives us information about the shape of the object.

To picture how that works, imagine an object on the floor of your living room, about a meter away, closely surrounded by a curtain that hides it from you. Your goal is to find out what's behind the curtain, and your only equipment is a bag of one hundred little marbles. All you can do is bombard the object with marbles, rolling them across the floor and under the curtain. They'll bounce off the object and come ricocheting out again. By looking carefully at how the marbles bounce and sending them toward the object at different velocities, you can get an impression of what the object looks like. If there's nothing on the floor behind the curtain but bread crumbs, then the marbles will roll straight over them and reappear on the other side as if nothing had happened. But if there's a sheet of wood at a 45-degree angle, the marbles will rebound exactly to the left or to the right. Those are both straightforward examples. But if you have to find out whether there's a thin wooden partition or a thick sheet of iron behind the curtain, the job gets a little harder. And just imagine shooting marbles to figure out whether there's a stuffed Mickey Mouse or Donald Duck doll behind the curtain. That's more challenging—extremely difficult, in fact—but not impossible! What you would need are (1) your marbles, (2) some idea of how Donald and Mickey make marbles rebound differently, and (3) a way of keeping track of the angles at which the marbles bounce back. This technique for looking at small objects has a fairly long history among scientists, who nowadays use particle accelerators to make the little bullets that they fire at objects. (I'll go on using

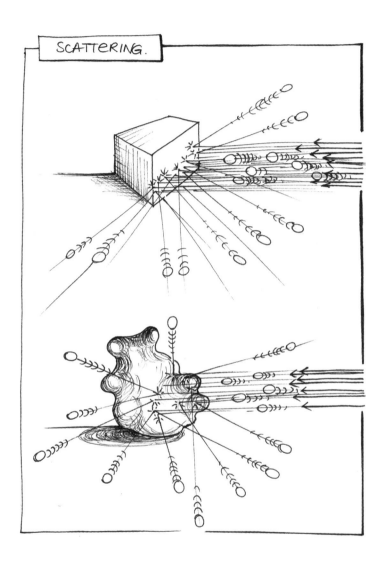

the word "bullet" here, but we're really talking about small particles.)

An electron microscope uses the very same method—rebounding particles—to create images of small objects like cells, details of an ant's eye, surfaces of a metal, or small structures in nanotechnology, so that we can study them. An electron microscope may use electrons, for instance, as projectiles to examine a surface or object. This technique has made it possible for scientists to go further into the micro-world than they can with conventional microscopes, ultimately to the point of unlocking the world of the atom.

There are three things we need to enter the world that remains hidden to our eyes and conventional microscopes:

1. Make tiny bullets and fire them—a particle accelerator.
2. Calculate how bullets are scattered by a particular shape—a theory.
3. Keep track of the scattered bullets—a detection device.

These are the three things with which my predecessors were equipped, and which we still use today in our search for ever smaller objects.

The particle-scattering trick is exactly what enabled the early twentieth-century scientist Ernest Rutherford to study the unique building blocks of each element, namely atoms, in detail. Later we'll see in more detail how he did that, but essentially, he fired small particles at a layer of gold atoms at high velocity and watched how they bounced. He had various ideas about the structure of atoms, but only one of them turned out to fit the measurements he and his assistants had made. They found that the atom was made up of a tiny, heavy, electrically charged particle, the nucleus, with a number of light electrons orbiting it. This new technique taught us that the world at its smallest scale had many secrets that would change science forever. That discovery set off an experimental gold rush, as scientists went searching for still more secrets of that mysterious world.

At this point I should mention that the traditional view of a particle as a hard little ball is not entirely correct. It's a bit more complicated than that. The rules of play for matter on that scale—quantum mechanics—state that particles also behave like waves. That sounds implausible, since the matter around us doesn't act like a wave at all. Yet experiment after experiment has shown conclusively that it is the case. Is that weird? Yes, very weird. It's good to realize that physicists think it's strange too. The only difference is that they're more used to the idea and have resigned themselves to the fact that not everything in the world of the atom is logical. One of those strange new rules states that the wavelength of a particle (and therefore its effective size) depends on its energy: the faster the particle moves, the greater its energy, and the smaller it effectively is. If we want the "bullets" we shoot from our particle accelerator to be sensitive to certain details of our object of study, then we need to use projectiles smaller than the structures we're investigating. By increasing the energy of the particles, we can make them smaller and smaller, so that they'll ricochet off even tinier objects and allow us to recognize even more details. The particle accelerator's job is to boost the energy of the particles as much as possible, so that they're as small as can be by the time we fire them at the object we're studying.

Like conventional microscopes, these "particle microscopes" have come a long way since the early days and are now used daily in many branches of science to reveal the surfaces of cells or materials. Any photos you see of nanostructures, the eyes of an ant, red blood cells, or cancer cells, for instance, were made with this technology. So the world's most powerful microscope is whatever particle accelerator can boost particles to the highest energy. Right now, that's the proton accelerator at CERN in Geneva: the Large Hadron Collider. The particles used in Geneva are at such high energy that their wavelength is approximately 10^{-20} meter, about one millionth the size of an atomic nucleus.

Although we often think of the particle accelerator as scientific equipment, it plays a larger role in our society. The best-known use of a particle accelerator, outside of science, used to be in an old-fashioned television, the kind with a screen that bulges outward. That screen is lit by a beam of electrons fired at high velocity, and magnets make the beam slither back and forth, hitting every point on the screen to create an image. The rise of LCD televisions has made that example awfully dated, but fortunately, particle accelerators also have a broad range of applications in health care and industry.

The ability of particle accelerators to make small objects visible has led to major scientific breakthroughs, because we can now produce detailed images of objects as small as blood cells or parts of ordinary cells. And for the past twenty years, manufacturers have been racing to build ever smaller structures, especially on computer chips. After a new idea or technique has been invented, it's important to examine the results (and any potential problems) in detail. That's why almost all high-tech companies use electron microscopes to produce images of the metal surfaces or nanostructures that they've made. The applications in health care are even more fascinating, because they're less familiar, so it's worthwhile to look more closely at a few of them. Even though these particle accelerators are, of course, not nearly as powerful as CERN's Large Hadron Collider, new developments in acceleration and detection technology in the scientific community have also benefited industry and hospitals.

When I'm giving a talk and I say that every modern hospital has a couple of particle accelerators, people look at me in astonishment, because few people realize that to make an ordinary X-ray or treat tumors, you just can't do without a particle accelerator. Even though, sad to say, many people have first-hand experience of X-ray images or radiation therapy for tumors, hardly anyone knows how those technologies work. Patients have other things to worry about, and when you're lying on that table, you don't ask exactly how it works. But because these are applications of particle accelerators, I'd like to tell you more about them.

The first well-known medical magic trick is looking straight through your skin to see your bones. Making an X-ray image is a lot like bouncing small particles off objects with a particle accelerator. The doctor fires something at you—in this case, X-rays (which are really just waves of light)—and examines whether they *did* or *didn't* pass through your body. Those rays have enough energy to go straight through soft skin and flesh, but *not* enough to penetrate hard bones. As soon as they hit bone or another hard material, they come to a stop. So in this case the main question is not the angle at which X-rays rebound, but whether or not they are blocked on their way through your body. Even though you can't see the rays that pass through the body with your naked eye, you can make them visible in a different way. It turns out that some materials turn dark when hit by X-rays (like the film in an old-fashioned camera, which turns dark when exposed to visible light). If you put a plate of this type of material behind the patient being scanned, the result is a detailed photo showing the exact locations of hard objects in the body. Of course, it also shows you where the hard parts are missing—for instance, if a bone is broken. You might compare it to the silhouette on a wall after a passing truck splashes mud all over you. That too is a negative image, showing where you stopped the mud from traveling onward. A brilliant idea, but where do those X-rays come from? That's where the particle accelerator comes into it. It whips up electrons to a high velocity and then fires them at a metal plate, releasing X-rays, which are aimed at the area the doctor wants to photograph. So without particle accelerators, there could be no X-ray images.

A traditional X-ray shows your bones in considerable detail, but sometimes we will want a sharper focus. One possibility is simply to fire more rays, but that does pose certain risks. There's a good reason that the professionals leave the room during an X-ray in the hospital or at the dentist's office: the radiation is hazardous. As it passes through your body, it can do serious cell damage and possibly even cause cancer. To limit that risk, we all accept that the image won't be quite as sharp as it theoretically could be. But there's a second way of obtaining a sharper image, namely by making the photographic plate more sensitive. This is like increasing the number of pixels in a digital camera. Scientists, like camera designers, are constantly working to improve their detection techniques. They've been very successful, and there are various initiatives to introduce the new technology in hospitals. After all, whenever we make our detection equipment more sensitive, we can take a better X-ray with the same amount of radiation—or we can take an X-ray with the same resolution while using less radiation. This doesn't make much difference if you're only taking one X-ray, but a CT (computed tomography) scan is the equivalent of two hundred X-rays. In that case, better technology means much less radiation and lower health risks for patients.

A good example of a joint initiative by science and industry—one that focuses on medical applications, detection techniques, and read-out chips—is the MediPix project, which has opened up possibilities for the color X-ray. The X-ray images we've discussed so far are always black and white, showing whether or not rays with a certain energy passed through the body. One disadvantage of black and white is that you can't see whether the rays were stopped by a bone five centimeters thick or a thin, two-millimeter iron plate. But suppose that instead of a single energy level, we could work with a whole spectrum of rays? That would give us more information, because we could see how much energy the particles need to pass through each obstacle. This principle has already been proven by researchers like Enrico Schioppa in his doctoral research at Nikhef, the Dutch National Institute for Subatomic Physics in Amsterdam. Dutch particle physicists perform a large range of experiments, as well as think about how they can contribute to new technologies. For a research and development group like Enrico's, seeing their research applied in an actual hospital would be a huge breakthrough. This requires a phenomenal amount of number-crunching, but fortunately, the Science Park in Amsterdam, where Nikhef is located, is also home to the Centrum Wiskunde en Informatica (CWI), the renowned Dutch research institute for mathematics and computer science. Together, these institutes are working on a test setup to show the world that a full-scale color X-ray system is possible.

Unfortunately, almost everyone knows somebody with cancer who needs radiation therapy. That type of treatment uses high-energy radiation, like X-rays but with much more energy. It turns one of the main drawbacks of that type of radiation, namely cell damage, into an advantage, by using it to destroy cancer cells. "Irradiation" means nothing more than being bombarded with energetic light rays that destroy cells. The particle accelerator comes into play because it is what generates the radiation. It's a lot like producing X-rays: ordinary particles are first accelerated to a high-energy state and then fired at a plate, releasing radiation that can be focused on the patient. Without a particle accelerator, there's no radiation. That's why every major hospital in the Netherlands has a fairly powerful particle accelerator, known as a cyclotron.

The good thing about radiation therapy is that, in the place where the tumor is located, it does destroy the DNA in the cancer cells. But it's no secret that on its way to the tumor, the radiation can also destroy or damage a lot of healthy cells, possibly even turning them into cancer cells. We'd like to keep that to an absolute minimum, for obvious reasons. Since our many experiments over the past decades have taught us how the particles lose energy in living tissue, we know that a great deal of damage is done in the area around the tumor. Doctors try to limit the damage to healthy cells by aiming particles at the patient from many different angles, so that the area right around the tumor receives the most radiation. But sadly, there is no way to completely prevent damage to healthy cells. That's a tragic fact for a patient with a tumor close to one eye, or near the spinal cord, but radiation therapy is also hazardous to children, because they have many cells that will go on dividing in the future.

One recent development that tackles these problems is called *hadron therapy*. It's a new kind of radiation treatment in which the entire dose is delivered to one place, so that it does the least possible damage to healthy tissue. Thanks to particle physics, we know how radiation moves through the body, losing more and more energy along the way. It's like a bicycle braking in the sand, which slows down fairly quickly and at an even pace, eventually coming to a stop. But we've also seen that there's another way. There are particles (protons) that pass through materials with almost no trouble at all. They slow down gradually but release their energy abruptly, all at once, when they are almost at a standstill. That sounds like just what we need: radiation therapy with protons instead of the usual photons, so that we destroy only the tumor instead of damaging all the healthy cells around it. If you know precisely how deep under the skin the tumor is found, then you can work out how fast to shoot the protons into the body so that they stop moving in that exact spot and do their damage there. And of course, you need a particle accelerator that gives you protons with exactly the right energy. Fortunately, that just happens to be the core business of particle physicists!

Right now, hadron therapy is still very expensive and available only in a couple of places. I'm no economist, so I can't calculate all the costs and benefits (and I don't want to get tangled up in a moral and ethical debate about the value of a human life, or of one extra year of it). But this is an amazing new possibility. In the years ahead, scientists will be looking to see what other innovations we can offer and how we can work together with doctors to achieve new successes.

In recent centuries, the drive to explain the things we see around us has led to a treasure trove of knowledge. Science keeps developing new techniques to overcome obstacles and delve further into the secrets of nature. In my own branch of physics, the particle accelerator proved to be the key to exploring the world of elementary particles. It allowed us to enter the world of the atom, study the atomic nucleus, and finally, with the discovery of the illustrious Higgs boson, travel to the deepest level we know of today.

2

The Atomic Revolution

In the early twentieth century, scientists were still far from realizing that empty space (the vacuum) wasn't empty but permeated by the elusive Higgs field. The world seemed orderly and well understood, and empty space was just empty. No one suspected the existence of the elementary particles we have identified since then, and when scientists spoke of basic building blocks, they were referring to the smallest bits of elements such as gold and oxygen. In those days, a new element was discovered just about every year. Scientists had even found a pattern in their properties and come up with a tidy architecture in which each element had its place: the famous periodic table of elements. This table seemed to have solved all their puzzles; chemists could happily spend their days placing substances with similar properties in neat rows and columns and tracking down the missing elements. It seemed only a matter of time before all the elements would be found.

While this was exciting, in a way, it was also a little tedious to check off boxes like an accountant. The world's scientists had no idea that they were on the brink of the greatest revolution yet. Much like the first humans to go into space, such as Russian cosmonaut Valentina Tereshkova, New Zealander Ernest Rutherford would lead the descent into the world of the atom. Rutherford would be the first to whom nature would reveal its secret forces and phenomena. These wonders, which were never before imagined by the physicists of his time, undermined completely the accepted scientific paradigm.

More than a hundred years ago, in 1912, Rutherford, an experimental physicist, launched modern particle physics with his groundbreaking experiment. He discovered that every atom has a very small central core, the nucleus, around which electrons revolve in fixed orbits. In the years that followed, the heroic efforts of other scientists made it possible to break down the tiny nucleus into smaller pieces. They found that the nucleus was composed of two distinct types of building blocks. Once the dust had settled—by the late 1930s—they realized that their tidy periodic table of more than eighty elements concealed a deeper level. To everyone's surprise, they learned that all the matter on earth, every atom of every element, and therefore all the matter in the universe, was composed of just three building blocks: protons, neutrons, and electrons.

This was when scientists began to construct a new logical framework that would make sense of all their strange observations. The new system that they uncovered overturned our perspective on nature. The pillars of this new architecture, such as quantum mechanics, are without exception strange and illogical, but they form the foundation of our physics to this day. And even though they still don't make intuitive sense to us, we've gradually grown accustomed to them and embraced the new logic of these illogical phenomena as a fundamental concept.

These steps into the atom and then the atomic nucleus formed a crucial phase in our search for the very smallest things, a phase that would eventually usher in a new concept of space and the matter it contains. It is fascinating to study the many different facets of this discovery: the technological breakthroughs, the researchers' persistence, their frustration when they ran into dead ends, their penetrating insights, and their euphoria when progress was made. But the most wonderful thing about those first steps is that their findings have now been applied in countless areas of our daily lives. This offers a simple response to the question frequently fired at fundamental scientists: "What use is it to us?" In fact, without the results and insights stemming from that fundamental research, our society would look very different.

Let's look at just a few examples from the vast range of findings and applications, moving straight on from the first fundamental layer, the atom, to the second layer, the nucleus. The atomic nucleus has shown us, among other things, how our sun is fueled, how we humans can build the ideal weapons for destroying us all (nuclear bombs), and—at the same time—how we may be able to tap into an inexhaustible source of clean, cheap energy.

This first step into the depths of matter revealed underlying patterns and provided a treasure trove of information about how our world really works. What we discovered about these minute particles compelled us to rewrite the laws of nature completely. The journey brought us the famous theory of quantum mechanics, elementary particles like protons and neutrons, the magical Pauli exclusion principle, the source of radioactivity, nuclear forces, nuclear fission and fusion, atomic bombs, and a potential solution to the global energy problem.

Many books could be—and have been—written about each of these subjects, but I'll stick to the essentials.

The Atom

The existence of sand, gold, water, and carbon on our planet is almost impossible to overlook. But you could easily remain unaware that the element neodymium exists, that water is actually a compound of oxygen and hydrogen, or that the air we breathe contains not only oxygen but also nitrogen. Around the year 1700, there were about ten known elements, and chemists were becoming ever better at isolating new elements. A century later, thirty-two elements had been discovered, and by the early twentieth century, almost all the chemical elements on earth had been identified. Once all the properties of each distinct element had been determined, scientists went looking for patterns in their similarities and differences. For example, silver and gold are different, but both are heavy metals and undeniably more similar to each other than to gaseous elements like oxygen or hydrogen. In the nineteenth century, the Russian chemist Dmitri Mendeleev classified atoms by weight; his system, known as the periodic table of elements, is still taught in every secondary school in the world. The great power of this system became clear when it was found to make accurate predictions. There were gaps in the table. And because the position of each gap provided information about the properties of the missing element, it was possible to predict not only the existence of elements that had never been observed, but also almost all their characteristic properties. Another major success of this classification system, or theory, is that not one element was discovered that did *not* fit into the system. By the early twentieth century, it was clear that there are just under one hundred elemental substances in nature, and the smallest distinct indivisible piece of each element was dubbed an atom. But until 1912, no one had any idea what an atom looked like.

The year 1905 was a productive time for Albert Einstein and a miracle year for physics in general. That was the year in which Einstein not only published his theory of relativity, but also explained that light is made of packages of energy—a theory for which he later won the Nobel Prize. He also introduced his now-famous formula $E = mc^2$, showing that mass and energy are equivalent. We'll run into that idea a few more times in this book, because it implies that you can convert mass into energy and vice versa. Those achievements were more than enough to earn Einstein a place among the scientific immortals, but he accomplished yet another extraordinary feat that same year: he calculated that there must really be such things as atoms. What Einstein showed was that to account for the erratic motion of particles in a fluid—for example, pollen drifting in the water—you have to assume that the grains of pollen are colliding with the building blocks of the fluid, the atoms, which have a particular distribution of velocities.

At that point it was already clear that an atom must be slightly smaller than a millionth of a millimeter, but scientists still had no idea what an atom looked like. They could not take a photograph of one, because there was no microscope sensitive enough. Just as Europeans used to fantasize about undiscovered regions of Africa or the Americas before the expeditions of the nineteenth century, and before 1959 we used to tell stories about what might be on the far side of the moon, early twentieth-century scientists indulged in speculation about the true nature of the atom. The idea that won the most support in the academic world at the time was that it was a sphere in which small, light, negatively charged electrons—which were already known to be inside the atom—circulated like currants in a kind of positively charged pudding.

The celebrated experimental technique used by Rutherford to determine the structure of the atom is essentially the same one we use today, more than a hundred years later, in our quest to find and understand the smallest building blocks. As I mentioned, the basic idea is that you can learn more about an object by studying how small projectiles bounce off that object. Rutherford's projectiles were what are known as alpha particles, moving at high velocity, which struck a layer of gold only a couple of atoms thick—like very thin aluminum foil, but made of gold.

Before the experiment, Rutherford had a clear idea about what would happen. If atoms really were like soft, fluid puddings with tiny currants moving around inside them, then the fast, heavy alpha particles would fly straight through them, the way a marble would fly straight through custard. To prove that he was right, Rutherford would have to show that all the alpha particles came out on the other side of the foil after he fired them. For that purpose, he used a screen that produced a small flash whenever it was struck by a helium nucleus (which is what an alpha particle actually is). As a precaution, he placed this screen around the entire experiment—not only behind the gold foil, but also along the sides, and even behind the source of the alpha particles.

This turned out to be a crucial choice. When he did the experiment (or actually, when he had his doctoral students Hans Geiger and Ernest Marsden do it for him—that's just how the scientific community works), they found that even though most of the alpha particles went straight through, once in a while—miracle of miracles—one would be deflected at a large angle. Occasionally, an alpha particle would even come bouncing almost straight back. This would be impossible, of course, if an atom were just a ball of soft pudding. Rutherford said it was the strangest thing he had ever seen, and made a vivid comparison: "It was almost as incredible as if you fired a fifteen-inch shell at a piece of tissue paper and it came back and hit you."

One possible explanation for this weird observation could be that there was something rock hard in the middle of the atom, hard enough to send a heavy, fast-moving alpha particle bouncing back. By systematically studying the angles at which the particles were deflected, they could learn more about that hard kernel. Ultimately, the only remaining explanation that fit all the data they had gathered was that almost all the atom's weight was packed into one tiny spot right in the center, with electrons revolving in regular orbits around it.

In traditional diagrams of an atom, the nucleus often looks fairly large. In reality, it is an incredibly tiny dot surrounded by a cloud of electrons. If you imagine that the cloud of electrons is as large as a cathedral, then the nucleus is the size of a fly on the altar, and the electrons drifting along the outer walls of the cathedral are smaller than bread crumbs. In other words, atoms, the building blocks of all substances on earth, are mostly empty. Rutherford's paper explaining the discovery is still very readable today, more than a century later. It's not easy to calculate the deflection (classical or otherwise) of a charged alpha particle by an atomic nucleus, but every university physics student makes the calculation at some stage, tracing the footsteps of the great master. The atomic model that resulted from this experiment is what most of us picture even now when we think of an atom, because it's what we learned at school:

- The nucleus is a small, charged ball that accounts for almost the entire mass of the atom;
- Lightweight electrons move in regular orbits around the nucleus; and
- Each electron orbit has a maximum number of electrons that it can accommodate.

From our perspective, this may not seem very shocking, but that's only because we were taught this model in school. It's important to realize how courageous it was for Rutherford to publish these findings. Even though his measurements proved that his model of the atom was correct, that model was completely incompatible with the laws of nature as they were then understood. Scientists' efforts to explain this anomaly ushered in an exciting time of many new insights, which in turn became pillars of today's natural sciences.

The motion of a small, light particle circling around something heavy is no great mystery to physicists. That's because this is just like the movement of a planet revolving around the sun, or the moon revolving around the earth. The laws of gravity were formulated more than three hundred years ago by Isaac Newton and have been applied very successfully ever since. We use them not only to describe the motion of the moon around the earth, but also to explain why an apple falls to the ground or where a cannonball will land. Although this theory, known as classical mechanics, was developed long ago, we still turn to it when we want to know how large objects will move.

But as successful as this theory has been, scientists ran into big trouble when they applied the same techniques to the movement of electrons in an atom. A different force is at work there, but in the most important respect (the strength decreases as the square of the distance between the particles increases), it behaves just like gravity. The problem is that an electron is electrically charged, and we know that electrically charged particles lose energy when they revolve around each other. This would imply that an electron loses energy, slows down, and crashes into the nucleus in a fraction of a second. At least, that's the theory. But in practice? Well, it would mean that atoms could never survive for long. And even if electrons *could* remain in perfect orbits around the nucleus for some inexplicable reason, why would they be confined to particular fixed distances, as Rutherford had found? The model of the atom that had emerged from Rutherford's experiments was therefore impossible, according to the accepted theories of his day, and raised more questions than it answered. In a classic confrontation like this one between theory and experiment, theory always loses out. A theory that says the *Titanic* is unsinkable doesn't last long once the ship is swallowed by the waves. Likewise, the observations of electrons calmly revolving in regular orbits made it clear that some essential insight was missing. But what?

It's questions like these, with no immediate answer, that keep physicists awake at night and lead to new steps in scientific thinking. An explanation like "That's just the way it is" doesn't satisfy scientists; their need to grasp the underlying logic drives them to experiment and debate their ideas endlessly, slowly working their way toward the solution. In the end, it was a Danish physicist, Niels Bohr, who saw the way forward in a flash of insight. He laid the first stone of a conceptual structure that could solve all these troubling questions in one fell swoop. Physics would never again be the same: quantum mechanics was born.

To understand how peculiar Bohr's proposal is, we have to draw a line between familiar physical phenomena and new concepts. Let's return for a moment to the comparison between electrons revolving around a nucleus and the moon revolving around the earth. Many of the principles are exactly the same, and yes, the laws describing the moon's motion around the earth are centuries old. But that doesn't mean they're easy to apply. Far from it. Figuring out approximately how long it will take a stone to fall from the top of the Empire State Building is fairly easy. But even something as seemingly straightforward as the motion of three planets cannot be predicted accurately without a computer. And that's not because we scientists are lazy; it's just one of many apparently simple problems we haven't been able to solve yet, despite all the mathematical tools at our disposal.

It's unfortunate that many of the physics problems presented in secondary school and university courses often require so much complex calculation. For many people, the mathematics is an insurmountable barrier. And even for university students and scientists, it's a distraction from the real point: the scientific concepts and ideas. At universities, we try to explain to our students that they should keep the physics separate from the mathematics, but that's easier said than done. Often the only way to get a firm grasp on the underlying world is through abstract mathematics. In dozens of situations, our intuition fails us, either because we can't think of the right analogy or because the logic of the new situation is fundamentally different from that of our everyday world. Returning to the analogy of the moon revolving around the earth, we'll soon see that we have to let go of our intuitions to grasp the solution to the problem of electrons in the atom.

In the case of the moon orbiting the earth, it's all fairly simple. The moon and the earth are both heavy, and just one force determines their motion with respect to each other: namely, gravity. The earth and moon exert a gravitational pull on each other. If the moon stood still, it would fall to the earth like a ripe apple. The only reason that doesn't happen is that the moon is in motion relative to the earth. Because it moves at such a great velocity, it would under normal circumstances escape the earth's gravity. But because these two effects—the moon rushing away and the earth pulling it in—hold each other in an astonishing balance, the moon makes neat circles around the earth and will go on doing so for another few billion years.

If you know a little more about the details of the law of gravity, or if you look back at your textbooks from secondary school, you'll soon realize that the velocity of an object in a fixed orbit around the earth depends only on its distance from the middle of the earth. As long as you know that distance, you also know its velocity. The closer it is to the earth, the faster it has to move to keep from crashing. To be exact, if it moves four times closer, it has to move twice as fast.

The International Space Station (ISS) circles the earth at a height of about 250 miles (400 kilometers). To avoid falling to the earth or being flung out into space, the astronauts on board have to maintain a velocity of approximately 17,000 miles (28,000 kilometers) per hour. Moving faster or slower is not an option. Although most satellites, like the International Space Station, are not very far away from the earth, a few of them deliberately orbit the earth at the same velocity at which the earth rotates. This special type of satellite, described as *geostationary*, completes one orbit every twenty-four hours so that it always remains above the same spot on earth. This is useful for espionage, among other purposes. If you do the math, you can see that this type of satellite has to be about 22,000 miles (36,000 kilometers) above the earth's surface in order to remain geostationary.

These laws of nature are universal. They apply equally well to planets orbiting the sun. So if you read somewhere that the planet Mars is one and a half times as far from the sun as the earth is, then you know right away that Mars moves about 25 percent slower than the earth. And because its orbit is also longer (it's farther away, so it has farther to go), it takes about 680 days to complete one circle around the sun, instead of the 365 days that the earth takes. The mass of the planet plays no role in this calculation, so you don't have to know what it is. This knowledge is not wizardry, but physics, and instead of memorizing facts about Mars, you can deduce many of them from universal laws. Very convenient.

Just as the moon circles the earth in a delicate balance, the electrons in an atom are in orbit around the nucleus. At first sight, the two systems are not so different, except that the nucleus and the electron are both electrically charged. That may seem unimportant, but in fact this difference is crucial. That's because gravity is no longer the only force at work; we must also consider electromagnetism. Comparing the strength of the two forces, we find that the electromagnetic force is stronger: about 50,000,000,000,000,000,000,000,000,000,000,000,000,000 times stronger. So gravity plays no significant role inside the atom. *Why* gravity is so weak compared to the other forces is still one of the great open questions of physics; we'll return to it in Chapter 6. For now, the interesting thing is that electromagnetic force, just like gravity, grows weaker the farther the particles are from each other. Electrons move faster the closer they are to the nucleus, the same way planets do when they're closer to the sun.

So far, the analogy seems to work perfectly. But that's an illusion. Here's the problem: while the moon and the earth do not lose energy as the moon orbits the earth, electrically charged particles do lose energy as they circle each other. In theory, this would slow down the electron slightly and send it crashing into the nucleus in a fraction of a second. In other words, the atomic model that emerged from Rutherford's experiments *could not exist*—or at least, not in theory. Yet the experiments showed that such atoms do exist in reality.

The stability of atoms remained a riddle for about five years before Bohr finally found the solution. Not everyone was happy with it, but it worked. His idea, which seemed very odd at the time—and honestly still does today—was that certain properties of particles at the atomic level cannot have just any value. Instead, their value has to be an exact (or integral) multiple of a certain minimum. Bohr had a specific property in mind that was well known to physicists: the angular momentum of the electrons in an atom.

Angular momentum is a characteristic relating to the energy of one body as it circles another. More precisely, it is the distance from the electron to the nucleus multiplied by the angular velocity. This is an important quantity for physicists because it is a quantity that does not change—that is, it is a *conserved* quantity. The best-known example of conservation of angular momentum is an event you've probably witnessed dozens of times: a figure skater speeding up as she brings her arms closer to her body during her final pirouette. By drawing her arms inward, she brings more mass toward her body, and because the angular momentum has to be conserved, she starts to spin faster. In a YouTube video, engineer Rolf Hut and his son demonstrate the same principle on rotating playground equipment.

Since the laws of nature are universal, they must apply to electrons moving in regular orbits around a nucleus. So their velocity depends only on their distance from the nucleus, like an astronaut's velocity as he orbits the earth. The angular momentum of the electrons depends on only one thing: their distance from the nucleus. Suppose you're an electron inside an atom. Once you decide on your distance from the nucleus, you have to move at one particular velocity (which determines your angular momentum). You're free to move in a larger orbit, in principle, but then you'd have to slow down a little. So far, so good. But now we come to Bohr's proposal for cracking the conundrum of the orbiting electrons:

> Niels Bohr: "The angular momentum of the electrons in an atom cannot have just any value. It must always be a multiple of the tiny quantity $\hbar$, better known as the reduced Planck constant."

In other words, angular momentum is a property that cannot have just any value. Instead, it comes in predetermined bits (quanta)—its value is *quantized*. If you think that's weird, you're absolutely right. So why did Bohr make this proposal, and how does it help?

We all know of one place where quantization is taken for granted: Legoland. When you use Legos to build something, its dimensions are always a multiple of the dimensions of a single Lego block. Similarly, you can't buy one and a half cartons of milk in the supermarket; you have to choose between one and two. But most properties, such as the speed at which you drive your car, are not restricted in that way. It would be a strange world indeed if everyone had to drive in multiples of 10 miles per hour. You wouldn't be allowed to go 65.7 or 72 miles per hour, but would have to stick to exactly 60, 70, or 80. Doesn't make much sense, right? Yet the quantization of some values is found not only in Legoland, but also in the world of the very smallest things.

There's no reason that the world of the tiniest building blocks should work the same way as our own. And Bohr's idea has clear advantages. If the angular momentum can take on only particular values (specifically, integral multiples of the minimum value), this would explain why electrons cannot spiral inward to the nucleus. Why? Well, imagine that an electron is in the innermost orbit and has an angular momentum of exactly one $\hbar$. According to the old laws of physics, it would lose energy and move inward. This would make its angular momentum slightly smaller: say, $0.98\hbar$. But that's not allowed! Why not? Because Niels Bohr's new law of nature says that the electron's angular momentum must be exactly $1\hbar$ or $2\hbar$, for instance, but not $0.98\hbar$—that's forbidden. So the electron may want to crash into the nucleus, but Bohr's new law won't let it. The electrons are trapped in their orbits, and the atom is stable, just as Rutherford observed in his experiment.

That may seem like cheating, because Niels Bohr never explained *why* it works that way, and as a matter of fact, we still don't know why. Quantization is still one of the great mysteries of physics. More generally, no law of nature has a logical explanation. Second-year university physics students always laugh at me when I ask why an apple falls to the ground. "Because the apple and the earth attract each other, obviously," they tell me. I know that's true, of course, but when I ask them *why*, they fall silent. It's fascinating: you can almost hear the penny drop (no pun intended) as they reflect on the deeper why of a phenomenon, some of them for the first time. The same discussion takes a minute longer when I teach master's students, who reply, "Because according to the general theory of relativity, space-time is curved by the mass of the earth." But if I go on to ask, "And why would the mass of the earth cause space to curve?" then silence descends. Quantum mechanics is exactly the same. It's one of the idiosyncrasies of nature that we've happened to notice. We accept that it's true and can use it in our calculations, but even scientists aren't sure *why*.

By calculating the distances at which the angular momentum of the electrons was exactly $1\hbar$ or $2\hbar$, Bohr could determine exactly what distances corresponded to stable electron orbits and how much energy the electrons in those orbits had. In other words, Bohr predicted not only the qualitative nature of electron orbits (known as shells) and energy levels, but even their exact quantitative values.

Let me invite the enthusiasts and connoisseurs among us behind the scenes for a moment for some extra information. What Bohr discovered, more precisely, was that for electrons in orbit n (where $n = 1, 2, 3, 4$, and so on):

$$distance = \frac{\hbar^2}{\alpha Q_e Q_p m_e} \cdot n^2 \qquad energy = \frac{\alpha^2 Q_e^2 Q_p^2 m_e}{2\hbar^2} \cdot \frac{1}{n^2}$$

Here m_e is the mass of the electron, Q_e and Q_p are the electrical charges of the electron and proton, α is the strength of the electromagnetic force, and $\hbar$ is the reduced Planck's constant. The only variable is n.

This is fantastic. It suddenly makes it clear why electrons in the second and third orbits are four and nine times as far away from the nucleus, respectively, as electrons in the first orbit, and have only one-fourth and one-ninth of the energy. Not only that, but because all the elements of this formula are known from other measurements, Bohr could also make an exact prediction. The smallest distance at which electrons orbit the nucleus ($n = 1$) turned out to be 5.3×10^{-11} meters, and those electrons are predicted to have an energy of -3.6 electron volts, just like the ones in the experiment. All the pieces of the puzzle fell into place. Scientists no longer had to say, "That's just how it is." Instead, they could predict those values from known constants of nature. Everything fell into place. Well, *almost* everything.

In short, Bohr discovered that when you shift your attention to the atomic level, you find the laws of nature operating in a fundamentally different way. By the way, he didn't conjure up the idea of quantization out of thin air. Another scientist, Max Planck, had previously noted that the only way to explain the amounts of radiation emitted by an object (such as the sun or a piping hot sausage) was to assume that the energy of the photons is a multiple of a particular small quantity of energy. That quantity was the same $\hbar$ that Bohr used for angular momentum. The reason *why* was still completely unknown, but in combination with Rutherford's observations of the atom, this discovery unleashed the quantum revolution. Max Planck and Niels Bohr both received the Nobel Prize—first Max Planck, in 1918, "in recognition of the services he rendered to the advancement of Physics by his discovery of energy quanta," and then Niels Bohr in 1922, "for his services in the investigation of the structure of atoms and of the radiation emanating from them."

The theory that describes this quantized world is called quantum mechanics, and a century later, it remains as mind-boggling as ever. In the years that followed, quantum theory and the related mathematical techniques were developed in greater detail. This provides insight into the strange way the elementary building blocks work at the level of the atom and is also crucial in biology and chemistry, at least if you want to understand how atoms absorb and radiate energy in the form of light. They do so in fixed units, because an electron always moves from one stable orbit directly to another one.

The predictions that follow from quantum calculations are often very strange and seem to defy all logic. Yet time after time, they turn out to be true. These predictions are many and varied: that particles can behave like waves, that they can be in more than one place at a time, that they can pass through normally impenetrable barriers, and other equally crazy stuff.

The classical view of an electron as a ball moving in circles around a nucleus went out the window a long time ago, with the introduction of wave mechanics. We now describe an electron as a probability cloud: a set of probabilities that the electron will be in a particular place at a particular moment. And quantum theory goes even further: the electron is actually at every point on the electron shell at once, and for every point, you can work out the chance of detecting the electron there. Every experiment imaginable has been carried out to verify or falsify quantum theory, but the theory has survived all these challenges and is now a cornerstone of modern physics. The situation remains awkward, because the origin of quantization is still unclear. But we cannot yet answer that question, so we'll just have to live with that—for now.

Of course, most objects we observe in our lives are much larger than an atom. So you might think that the subtle laws of quantum physics, which operate only at that level, are unlikely to affect our daily lives. But that would be a gross misunderstanding. For example, without quantum mechanics we would not understand how the components of a computer chip work, and I invite you to spend a day without using one of those. In other words, a day without your radio, thermostat, telephone, iPad, car, laptop, or digital camera, and without the Internet. I'm not suggesting that you give thanks to a whole list of physicists every time you take a photo with your phone, start up your computer, or use your car's navigation system, but it's

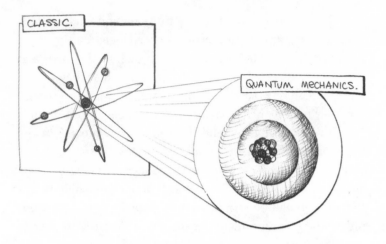

good to be aware that these highly fundamental discoveries have had a huge impact on our lives. Some have even suggested that quantum effects operate on a larger scale than we deem possible right now; if so, this could further our understanding of how the first life, or complex molecules, formed on earth.

Seeing a particle as a probability cloud that spreads through space like a mist and can pass through barriers impenetrable to classical particles—that was just the beginning. Many other secrets still lay hidden in the world of the atom.

Even if we accept that Niels Bohr is right and electrons move in fixed orbits, or shells, around the nucleus, that still doesn't solve all our problems. Why don't the three electrons of a simple atom like lithium all use the orbit closest to the nucleus? After all, that would require the least energy. The two electrons of the slightly smaller helium atom do manage to share the same orbit. So scientists were puzzled to see that the third electron in the element lithium is off on its own in a much more distant orbit, where it is much more weakly bound to the nucleus. Incidentally, this has practical consequences: lithium's chemical behavior is completely unlike that of helium. But that's not the main point here.

In secondary school chemistry, we learned *what* goes on inside the atom: each shell has its own maximum number of electrons (2, 8, and 18 electrons for the first, second, and third shells, respectively). But *why*? Beyond the laws of quantum mechanics, an additional, ad hoc law of nature was needed to shed some light on the *why* of this phenomenon. Wolfgang Pauli was the first to point the way out of this quandary, when he proposed:

> Wolfgang Pauli: "No two electrons in the atom may be identical."

If no two electrons may be identical, then we need to look at what makes an electron unique. What do you need to know to identify a particular electron inside an atom? Scientists call such properties quantum numbers. To identify a particular car, you can use the properties of make, model, engine capacity, and year; a person has characteristics like sex, age, weight, and height. For electrons in an atom, quantum mechanics seemed to predict that there were three properties by which an electron could be uniquely identified: n (which orbit the electron is in) and the more arcane properties l and m (which have to do with the exact paths followed by the electron). The important thing is that this model, in combination with Pauli's new constraint, states that only one electron can occupy the innermost orbit, namely the electron with the quantum numbers $(n,l,m) = (1,0,0)$.

But hold on a second! If you've been paying close attention, you realize that this contradicts the experiment that showed that there were two electrons whizzing around in the atom's innermost shell. How can these electrons be identical? Samuel Goudsmit and George Uhlenbeck, two young doctoral students in theoretical physics in Leiden—my current home and one of the centers of global physics research—had just found a way to distinguish between those two apparently identical electrons. They weren't absolutely the same after all, so they were allowed to whirl around together in the first orbit.

More specifically, Goudsmit and Uhlenbeck postulated that electrons had an additional, previously undiscovered

One of the wall formula murals in the inner city of Leiden commemorating the discovery in the 1920s of electron spin by two Leiden physics PhD students, George Uhlenbeck and Samuel Goudsmit. The mural is one of several in the wall formula project, a combined initiative by the artists Ben Walenkamp and Jan Willem Bruins from the TEGEN-BEELD foundation and physicists Sense Jan van der Molen and Ivo van Vulpen. The building on the left, "de Latijnse school" (Latin school), was once attended by Rembrandt van Rijn.

property, which they called "spin." Although the analogy does not fully hold, you might picture it as the rotation of the electron. They showed that electron spin comes in just two distinct varieties: electrons spin either to the right or to the left. There were no other flavors. Imagine a perfectly uniform ball spinning. You can't tell it's spinning *unless* you touch the ball. That was why the property of spin had gone unnoticed until then— although a year earlier, experimenters had observed a strange effect for which they had no explanation. The new property of spin discovered by the young scientists in Leiden soon provided the solution.

Electrons had always had this "hidden" property, of course. Similar discoveries happen in everyday life. Say you run into two forty-five-year-old men in a local bar, both about six feet tall and two hundred pounds. Both work at the same university and both vote the same way. Perhaps they are even both Dutch, and so they both like bread and butter with chocolate sprinkles. They may seem almost identical, and you might imagine they would be best friends since they have so many things in common. But you might change your mind completely when the conversation turns to soccer and you find out that one supports Sparta and the other Feyenoord—two rival Rotterdam teams. You would probably never have learned that important fact if you hadn't asked the right question. Electron spin, likewise, was the unexpected fact that changed everything, and Pauli saw right away that it was the property he needed.

Pauli's constraint, discussed earlier, is named after him: the Pauli exclusion principle. It successfully explains why each electron orbit in an atom can accommodate a different maximum number of atoms. The first orbit, for instance, holds electrons with the quantum numbers $n = 1$, $l = 0$, and $m = 0$. So it can hold only two electrons, one with a spin of $+\frac{1}{2}$ and one with a spin of $-\frac{1}{2}$. Those are the only two possible spin values for an electron. Any other electron would have the same quantum numbers as one of the two electrons already there. And that is no longer permitted, because of Pauli's new principle. The third electron would therefore be forced into the second stable orbit, further from the nucleus. And the possible values of the quantum numbers n, l, m, and s (spin) limit the number of electrons that can occupy that shell before it fills up.

I might add that Bohr's model of the atom is not as simple as the version I present here. The energies of the different shells overlap, and predicting the chemical properties of molecules is a very complex problem. Even though hardly anyone has heard of spin as a property of elementary particles, it is a fundamental concept of particle physics. All known elementary particles are divided into two camps. In the first are the *fermions* (spin = $\frac{1}{2}$), the building blocks of matter, and in the second are the *bosons* (spin = 1), also known as force carriers. There is only one particle with a spin of 0, which was not discovered until 2012. It is the main character of this book: the Higgs boson.

Even a property as strange and well-hidden as spin has practical uses. In medical science, an MRI scanner uses the spin of the nuclei of hydrogen atoms in your body, which can be detected with the help of a very strong magnetic field and an electromagnetic field. MRI scans are used to form pictures of body parts such as bones and cartilage. For example, an MRI scan of the knee of a particle physicist who has always been an avid amateur footballer may show that his cartilage is worn and that he should spend his time writing books instead of dreaming of a future on the FC Barcelona first team. Of course, I'm not the only one who is grateful for the technique, so it's not surprising that in 2003 Paul Lauterbur and Peter Mansfield won the Nobel Prize in Physiology or Medicine "for their discoveries concerning magnetic resonance imaging." Spin is also central to the development of quantum computing, in which a particle can be in two spin states ($+\frac{1}{2}$ and $-\frac{1}{2}$) simultaneously.

The Atomic Nucleus

By this stage it was clear that the elementary building blocks of all substances on earth all consist of atoms, and that each atom has a relatively heavy nucleus orbited by electrons. What remained a mystery was whether the nucleus was an indivisible speck or had a structure of its own. The idea that every nucleus is indivisible and unique may seem very logical, because the nucleus gives an element its characteristic properties. And why should it be made up of even smaller parts? But in the two decades following its discovery, scientists eventually found that far from being indivisible, the nucleus consists of two separate building blocks: an electrically charged proton and a neutral neutron. These two building blocks are about equal in weight and both much heavier than the electrons that orbit them (about two thousand times as heavy), and they're present in roughly equal amounts in the nucleus. Protons and neutrons make up the nucleus of every atom.

Notice that we are taking a giant step here, from one hundred distinct elements to just three building blocks. It could hardly be simpler. Absolutely everything—from water to gold, from coal to human brains, from a nuclear warhead to a duck-billed platypus, from a glass of beer to the Great Wall of China—is made up of just three building blocks: the proton, the neutron, and the electron. When the nucleons (in other words, the protons and neutrons) are combined differently, the atom's properties change.

Just as the structure of the atom had opened the door to a new world, the unraveling of the secrets of the nucleus revealed new phenomena at work, which even quantum mechanics could not entirely explain. I wish I could give you the full historical perspective, so that we could pay homage to all the physicists who contributed to these discoveries, but that would distract us from the essentials that everyone should know. So strap on your seven-league boots for a brief survey of how we explored the nucleus with the help of particle accelerators, and then of the complex puzzle we faced as we tried to explain what we had found there. The solutions—nucleons and two new nuclear forces—appeared out of nowhere, but they formed the foundation of particle physics around World War II. For almost all practical purposes, they are still recognized as the elementary building blocks of nature.

After Rutherford's success in studying atoms, you might think the logical step would be to repeat the same trick using a microscope with an even higher resolution—much higher if necessary. Rutherford's particle accelerator was not yet strong enough to probe the nucleus, but of course, technology did not stand still. The size of the smallest structures that a microscope can observe is directly linked to the size of the particles you fire at them. And since quantum mechanics had taught us that we could make particles smaller by accelerating them to a higher energy, that was our constant quest. In spite of the tiny size of the nucleus, we would eventually reach the high energy required and shoot it to pieces.

The first step in the search for smaller structures was taken by Rutherford in 1917. The smallest type of nucleus we knew about then was the nucleus of a hydrogen atom: a single proton. By bombarding all kinds of materials with high-energy particles, Rutherford discovered that protons were also hidden inside the nuclei of other elements. He reached that conclusion because he saw that in addition to the alpha particles he was firing, the collisions released protons. He knew they must have come from inside the bombarded nuclei. That seemed logical, because if the nucleus contained one proton for every electron spinning around it, that would explain why atoms had a neutral electrical charge—an experimental fact.

Now we turn the spotlight on a little-known Dutchman who found an important piece of the puzzle: Antonius van den Broek. In a letter to the editors of the science journal in which Rutherford had informed the world of his findings about the makeup of the atom, Van den Broek noted that the charge of the nucleus was probably exactly equal to the number of electrons. That implies that the charge of an element's atomic nucleus determines its place in the periodic table. Although you won't find Van den Broek's name on most lists of great physicists, his insight deserves to be mentioned here.

One little problem, though. It became clear that the nucleus was twice as heavy as it should be if it were made up entirely of protons. In 1932, James Chadwick pushed his ingenuity to the limit and discovered the existence of the neutron: a particle with a neutral electrical charge that weighed about the same amount as the proton and was also in the nucleus. This was the last puzzle piece. Each atom had three components: protons and neutrons (in more or less equal numbers) tightly packed into the nucleus, and electrons whirling around them, making the net atomic charge neutral. The number of protons determines what type of atom we're dealing with. One simple way to picture the nucleus is as an object made of little balls:

the protons and neutrons. Most physicists probably think of it that way—at least, I know I do—and it turns out to be a fairly accurate image.

If every nucleus really is composed of those two particles, then we had a big problem. How could a nucleus of that kind exist in the first place? According to electromagnetic theory, two positively charged particles should push each other away with great force. So how could all those protons be packed into the nucleus? And what about the neutrally charged neutrons? What kept them in place? These were valid questions that physicists could not yet answer. To make matters worse, no one was able to solve the problem by modifying known laws —the approach that had worked so well in the past, leading to new principles like quantum mechanics and the theory of relativity. No matter how attached you are to your image of the world, if the experimental facts tell you you're wrong, then you have to do something. Back then, the theory said that a certain force (electromagnetic repulsion) would push apart the protons in the nucleus, but that didn't happen. So there was apparently another force at work, a force even more powerful than electromagnetism, holding together the particles in the nucleus: the strong nuclear force.

You might compare the situation to a marriage in which the partners are tired of each other and argue all the time but stay together because their love for their children is stronger. Or you might think of two politicians from different parties who have to work together to achieve a common goal. Like political intrigues, the dynamics of the laws that dictate how nucleons are held together are complex and fascinating. They involve a new force of nature—both powerful and strange, because even though within the nucleus the new force is stronger than the electromagnetic force, at slightly greater distances it quickly becomes less powerful. It's a good thing, too, because otherwise all the nucleons on earth would merge into one huge nucleus, and we couldn't even exist. The discovery of the nuclear force not only led to interesting insights, but also made it clear that enormous energy lay hidden in the nucleus.

When we study the nucleus, a traditional microscope is not the right approach. We saw earlier that if you want to know what's inside a walnut, the tool you need is not a microscope, but a hammer. When we study the atomic nucleus, we run into a similar problem. You might compare it to studying an old-fashioned dial alarm clock to figure out exactly how it works. However closely you look at the outside, you won't learn a thing about the clockwork that moves the hands. There's just one way to find out how that works: unscrew the screws and open up the clock. But we don't have a screwdriver for the atomic nucleus, so that leaves us with only one option: smash the clock and see what pieces you end up with. If you do that carefully enough, and often enough, you can reconstruct the inside of an alarm clock.

This destructive technique is exactly what scientists used. Our particle accelerators grew more and more powerful, and we kept bombarding atomic nuclei to see what would come out of them. It wasn't easy. Both the projectiles and the nucleus were electrically charged, and as the distance between the particles decreased, the repulsive force between the particles increased rapidly. In other words, the closer a charged projectile particle gets to the nucleus, the harder it is pushed away. The only way to reach the nucleus is by using an enormous amount of energy.

In the early 1930s, we started to make swift progress. In Cambridge, the scientists John Cockroft and Ernest Walton designed a particle accelerator that could shoot particles with enough energy to reach the atomic nucleus. For this world-class experimental achievement, known as splitting the atom (nuclear fission), they won a Nobel Prize. But their rivals were not far behind. That same year, Ernest Lawrence pulled off the same feat in the United States, using a much simpler device that he had designed, the cyclotron. This won him a Nobel Prize of his own. While the particle accelerators in England were the size of a house and cost millions of pounds, Lawrence's particle accelerator was the size of a human hand and cost only twenty-five dollars! This is another great example of how, in science, clever ideas count for more than brute force. Cyclotron technology still forms the basis of all particle accelerators, and every modern hospital now has a cyclotron, which does the hard work of producing radiation for cancer treatment.

In the early 1930s, it quickly became clear that the arrangement of the protons and neutrons in the nucleus tells us something about the strength of the force that holds those two kinds of particles together—and therefore about the stability of the nucleus (table 1). The rule of thumb is that protons and neutrons are found in the nucleus in about equal numbers. Helium has two protons and two neutrons; oxygen has eight protons and eight neutrons. But as we move from lighter to heavier elements, they have more and more neutrons relative to the number of protons. Gold, for instance, has 79 protons and 118 neutrons. The most important things to keep in mind are that the number of protons tells us which element we're dealing with, and that only nuclei with the right mix of neutrons and protons can survive for long.

   If you take an atom of gold with an ordinary nucleus and add a neutron, it will still be gold (because the number of

Table 1. Several Stable and Unstable Elements

Element	Number of protons	Number of neutrons	Form	Stability
Helium	2	2	Gas	Stable
Beryllium	3	3	Metal	Stable
Oxygen	8	8	Gas	Stable
Gold	79	118	Metal	Stable
Gold	79	119	Metal	Unstable

protons is still 79), but the extra neutron will make the atom slightly heavier. This is called a new *isotope* of gold. But the heavy nucleus turns out to be unstable. That means the arrangement of protons and neutrons doesn't work well. By rearranging them and/or emitting radiation, the nucleus can create a much better situation, energetically speaking. And that's just what it does. We call an unstable nucleus radioactive, and the radiation it emits is called nuclear radiation.

In the case of our gold atom with an extra neutron, it takes an average of two seconds before the nucleus transforms (or "decays"). In this particular case, the extra neutron is transformed into a proton and an electron (and a neutrino, but we'll talk about that later). The electron goes flying off at high velocity: we call that a beta ray.

You now know enough to see what happens to the atom after it emits the beta ray. We started with an atom of gold with 79 protons and 119 neutrons. As soon as one of the neutrons turns into a proton, we have a nucleus with 80 protons and 118 neutrons. So even though the atom is still just as heavy (since the number of nucleons is still 198), it has changed in character. You see, an atom with 80 protons in the nucleus is an atom of a different element, namely mercury. Gold has transformed into mercury! Even after radioactive decay, the atom is sometimes still unstable. In that case, it goes through radioactive decay again. This decay chain continues until the result is a stable atom.

Over the years, scientists have studied all the isotopes of all the elements, looking at which nuclei are stable, which are not, and how quickly the decay takes place on average (the half-life of the isotope). The type of radiation emitted depends on the type of decay. There turn out to be three kinds: alpha, beta, and gamma (table 2).

Along with these three types of radiation, depending on the kind of decay, unstable atoms also have a very broad range of half-lives. An unusual gold atom like the one in our example decays after two seconds on average, but a uranium nucleus tends to decay only after a few hundred million years (by breaking almost in half). So after a billion years, a chunk of uranium is still highly radioactive. This is exactly what makes uranium such a hazardous form of nuclear waste.

Even before we knew all this, nuclear radiation was a familiar phenomenon. Take, for example, the radium that used

Table 2. The Three Types of Radiation

Alpha radiation	Emits a helium nucleus	Atomic number goes down by two	Mostly harmless
Beta radiation	Emits an electron	Neutron becomes a proton	Slightly harmful
Gamma radiation	Emits a photon	Nucleons rearrange	Very harmful

to be in the glow-in-the-dark paint on a watch face or the radiation that reveals the inside of a hand (X-rays, a type of photon). But after the atomic revolution, we finally understood the origin of radiation and how we could use it to learn more about the nucleus. Radioactivity has a bad reputation, but it's also used in hospitals—not only to irradiate cancer cells, but also to look for blocked arteries. A small amount of radioactive material, called a tracer, is added to the bloodstream and spreads throughout the body. By measuring the radiation that's emitted in the blood and then comes out of the body, you can produce an exact image of the cardiovascular system. You've probably seen images of that kind. How else could a doctor figure out whether there's a blockage, and where it is? So despite the bad press, radioactivity comes in handy. Like physicists.

The extent to which neutrons and protons attract each other in the nucleus, known as binding energy, depends on the combinations of protons and neutrons. This became clear fairly quickly once scientists were finally able to make nuclei collide. They found that the binding energy has a crucial property that makes it possible even for stable nuclei to fuse or split apart. Furthermore, those processes produce energy. Before we go into all the implications of this discovery, let's consider an analogy that connects fusion and fission to our everyday world.

Countless examples clearly illustrate that when a company experiences a lot of growth, it becomes more efficient at some point to split it into smaller units. The advantages of remaining unified no longer outweigh the advantage in flexibility and energy that smaller units can offer. The costs and the legal complications surrounding these reorganizations can quickly be recouped. At the other end of the spectrum, the rules are different. While large companies tend to become more profitable after a split, smaller companies are better off merging. There are considerable up-front costs involved in making mergers happen, but the investment leads to new energy and greater profits.

We see the same patterns in the atomic nucleus. When large nuclei are thrown off balance—for instance, because we fired an extra neutron at them—they sometimes split into smaller pieces (two new elements). The new configuration, in which the nucleons are divided over the two pieces, is much more energy efficient than if all of the nucleons had stuck together loyally in a single nucleus. Because of this better distribution of nucleons, the combined weight of the two pieces is less than that of the original nucleus before it fell apart. It may seem strange, but it's true: mass actually disappears. In nuclear plants, we take advantage of that fact to produce energy by splitting uranium nuclei.

Surprisingly, the nuclei of light atoms show the opposite pattern. They fuse, and in the process of forming a heavier nucleus, they too release energy. The more efficient arrangement of nucleons makes the mass of the new, larger nucleus less than the combined mass of the two smaller, separate nuclei. Again, strange but true. This last insight is what finally revealed to us what fuels the sun: namely, the fusion of small hydrogen nuclei into somewhat larger helium nuclei. Some of the energy released in the process reaches us here on earth as light and warmth. That's no trivial insight. A hundred years ago, no one in the world knew what gave the sun the energy to keep burning. But nowadays we have such a thorough grasp of these fusion processes that we can even use them here on earth. Creating a "sun on earth" is the essence of the incredibly powerful hydrogen bomb, with which the human race could easily wipe itself out. At the same time, nuclear fusion may also be the ultimate solution to the world's energy problem—as long as we can keep it under control.

Atoms were actively split for the first time in 1938, in Germany, by Otto Hahn, Lise Meitner, and their colleagues. They fired neutrons at the atomic nuclei of the heaviest naturally occurring element on earth, uranium. Not many people would have thought of doing that, but physicists have experimentation in their blood. It was convenient to use neutrons because they have no electrical charge, so they aren't pushed away by the positively charged nucleus. Hahn and his colleagues expected to create a somewhat larger nucleus, with a single extra neutron, but to their surprise they saw that the process created lighter elements instead. It seemed that they had succeeded in creating a new, larger nucleus, but it was tremendously unstable, and it therefore made energetic sense for that new nucleus to split into smaller pieces. Even more unexpectedly, the weight of all the pieces added together was less than the weight of the original uranium nucleus plus the neutron fired at it.

But how was that possible? After all, if you cut a piece of cheese in two, the total weight of the two halves is equal to that of the original piece, isn't it? While that's certainly true in the case of cheese, atomic nuclei work differently: if the arrangement of the nucleons becomes more efficient, it takes less energy to hold them together, so the new particles may weigh less in total than the original one. This lost mass is converted into energy according to the well-known formula $E = mc^2$. And even though the difference per nucleus is only a fraction of the mass of a proton, the conversion factor between mass and energy is so large that this corresponds to a huge amount of energy.

So if you give a uranium nucleus a little shove by firing a neutron at it, it splits and there's a net increase in energy. Some of the energy that was always stored in the nucleus has been released. A net increase in energy sounds like a good investment, to stick with our business analogy. One important and interesting observation was that the decay of the uranium nucleus into two smaller nuclei left a few free neutrons. These "free agents" did not become part of a nucleus, but struck out on their own.

That seemingly unimportant detail, which is nonetheless a fundamental scientific fact, has had an unbelievable impact on world politics. That's because those free neutrons can make another uranium nucleus unstable, so that it falls apart: a chain reaction. This observation was the start of the great nuclear arms race. Three years after this measurement, the first nuclear reactor had been built, and another two years later, the first nuclear bomb exploded over the city of Hiroshima in Japan. In business terms, this was a very short time to market for fundamental physics. The results were horrific.

When a uranium nucleus is split, the reaction produces more free neutrons than the one that was fired at the nucleus. If the energy level of these neutrons is just right, then they too can bump into other uranium nuclei and cause them to split, releasing still more neutrons that bump into still more nuclei, and so on and so forth. The process continues until the uranium is used up. That sounds harmless enough, and at first sight, it actually looks ideal. If exactly one neutron is produced per decay, then it's like a series of dominoes that all fall over after the first one is pushed. So you can "burn up" a whole chunk of uranium metal just by firing one neutron into it to break apart the first nucleus. The rest of the process is automatic and produces "free" energy as the uranium nuclei decay one by one. You could use that energy to evaporate water, and with the steam you produce, you can generate electricity—just the same way you would in an ordinary coal or gas-fired power plant.

It was the famous scientist Enrico Fermi in the United States who created the first nuclear reactor. This was soon after he received the 1938 Nobel Prize for his research on radioactive substances with the help of slow neutrons. All he had to do to make that first nuclear reactor was bring a whole lot of uranium together in one place. That was difficult, but not impossible. One problem was that the neutrons produced by uranium decay escaped just a bit too fast to cause the decay of other uranium nuclei. By slowing down these neutrons a little, he could use them more efficiently. Just as people walk more slowly in a swimming pool, neutrons can be slowed down by materials like water and graphite (a form of carbon). Fermi concluded that piling up a lot of pallets of uranium, alternated with blocks of graphite, should do the trick. The uranium was gathered together in secret and stacked, and on December 2, 1942, a successful test took place. Fermi and his team had built the world's first nuclear reactor, Chicago Pile-1.

Now, seventy years later, there are lots of nuclear reactors all over the world, and although there are different varieties, the principle is always the same. They burn radioactive fuel (uranium-235, plutonium, or thorium), heat water, and use steam to generate electricity. On the face of it, nuclear energy seems ideal: it doesn't emit any carbon dioxide, and with just a small amount of fuel you can produce a lot of energy for a long time. This last advantage is the reason it's used in nuclear submarines. But as we all know, there are also disadvantages. You can't touch radioactive materials, because the radiation is so unhealthy. And once the fuel has been used up, you're left with the waste: the smaller nuclei, which are highly radio-active and have to be stored securely for thousands of years. Recently, thorium reactors have been getting a lot of attention. The thing is, we have a lot more thorium here on earth than uranium, which is scarce. Thorium also produces hundreds of times less radioactive waste and—not insignificantly—you can't use it to build atomic bombs.

The disasters in Chernobyl and Fukushima are grim reminders that, without the right safety measures, nuclear plants can lead to serious trouble. With that in mind, the choice of location for the world's first nuclear reactor is quite surprising: underneath the viewing stands of the deserted football stadium Stagg Field in Chicago. If the experiment had gone wrong, the disaster would have been beyond imagining. And it could easily have turned out that way.

A row of falling dominoes—the analogy sounds innocent enough, but if more than one neutron is released each time an atom splits, then the chain reaction can quickly get out of hand. Every step of the way, the number of splitting uranium nuclei increases. If two neutrons are released instead of one, for example, then the number of uranium nuclei that decay in the first five steps is 1, 2, 4, 8, and 16, in that order, and by step 50, the number is 1,125,899,906,842,624. If each of these steps happens very fast—in a fraction of a second, say—then nuclear fission can be used to make a bomb. When research by scientists in Paris showed that each uranium-235 nucleus produces not one but three neutrons at every step, and that the chain reaction could go very fast, the Hungarian scientist Leó Szilárd, living in the United States, started to panic. He asked the scientists not to publish their results, fearing that it would give the Germans the same idea it had given him: an atomic bomb.

They published anyway. This all happened not long before World War II. Szilárd, in cooperation with a few other scientists, managed to convince Albert Einstein to sign a letter to President Roosevelt pointing out the dangers of the potential new bomb. The U.S. government, seeing the urgency of the problem, eventually supplied Fermi with the uranium he needed for his first nuclear reactor and started a uranium enrichment program.

There's a fair bit of uranium in nature, but most of it is uranium-238 (with 92 protons and 146 neutrons). For nuclear fission with extra neutrons, you need uranium-235, an isotope with three fewer neutrons in the nucleus. That can be found in nature too, but unfortunately, it's a tiny fraction (0.7 percent) hidden away in the "ordinary" uranium. To build a nuclear reactor, you could leave the uranium-235 mixed into the ordinary uranium. But to make an atomic bomb, scientists needed a chunk of fairly pure uranium-235, so that (in principle) each neutron would split another nucleus. And a small chunk wouldn't do the trick, because the nuclei on the surface send their extra neutrons out into the air instead of causing the decay of even more nuclei. The larger the chunk of uranium-235, the smaller the fraction of neutrons that escape. At a certain weight, known as the *critical mass*, the chunk will spontaneously explode. It's not a huge weight, I might add: about 50 kilograms of uranium-235 (a ball with a radius of 10 cm), or just 10 kilograms of plutonium-239. In each case, the chunk is about the size of an orange.

So basically, here's how an atomic bomb works: you put together a chunk of uranium-235 or plutonium that has at least the critical mass. The usual approach is to make that chunk by shooting two pieces (each under the critical mass) at each other at high velocity, immediately creating a chunk large enough to explode spontaneously. If you do that too slowly, the two smaller pieces burn up and then fizzle out, and nothing happens. So "all" the bomb-builders needed to do was get their hands on 50 kilograms of pure uranium-235 or 10 kilograms of plutonium, aim, and fire.

Separating the two isotopes of uranium, or at least making a chunk of uranium with a large enough proportion of uranium-235, is called *enrichment*, and it's very difficult, because the two isotopes are chemically almost identical. They're often separated with large centrifuges, which are a lot like washing machines. By spinning around uranium gas at high velocity, they draw a little more of the heavier uranium-238 toward the outer wall, with purer uranium-235 remaining in the center. This enrichment process is the greatest barrier preventing new states from developing atomic bombs. Lucky for us. The depleted uranium that remains is used in ammunition, because its weight enables it to pierce armor plating easily.

But you can also build an atomic bomb without uranium, by using plutonium. In principle, you need a lot less of that. The problem is that plutonium doesn't occur in nature, but has to be made in a nuclear reactor. So it's very difficult to come by. During World War II, however, the pressure was extreme enough to make it happen. The first side to build the bomb would win the war, and in that all-or-nothing race, a mind-boggling sum of money was pumped into the project. Thanks to that same sense of urgency, a long line of top scientists spent a few years working on the Manhattan Project to develop the new weapon, and on July 16, 1945, the first atomic bomb was successfully detonated in a desert in New Mexico, an event known as the Trinity test. The Americans were the first to succeed in making each type of bomb (uranium-235 and plutonium), and both were used in the terrible bombings of Hiroshima and Nagasaki on August 6 and 9, 1945, respectively.

To anyone interested in that hectic five years from the discovery of nuclear fission to the building of the first atomic bomb, I strongly recommend Richard Rhodes's *The Making of the Atomic Bomb*. This detailed account describes the physics, technology, and politics of the "race for the bomb" during the first few years after the discovery of nuclear fission—a discovery that opened Pandora's box. The knowledge released into the world can never be put back in the box again, so now it's up to politicians to make sure the human race doesn't wipe itself out.

While it's more energy efficient for heavy nuclei to split (after a little push), it's more energy efficient for lightweight atoms to fuse into larger nuclei. This creates a new, heavier nucleus, which is lighter than the total weight of the two separate nuclei. Somehow it takes less energy to hold all the nucleons together in a large nucleus—in other words, the binding energy for the arrangement of particles is less. That makes the nucleus lighter (as predicted by the world-famous formula $E = mc^2$, which describes the relationship between energy and mass). Of course, it's still strange that the two nuclei merge into something lighter than their total weight.

Today we know enough to reason it all out, but arriving at that understanding was far from easy. To discover and reproduce this phenomenon, you first have to bring the two lightweight nuclei into contact. That's tricky, because both are electrically charged, so they repel each other, like two magnets when you try to press the "wrong sides" together. You have to push pretty hard to make it work. So there was no way to observe this phenomenon until someone built a particle accelerator strong enough to make the lightweight nuclei move fast enough to collide. That fortunate scientist was Mark Oliphant in Cambridge, who demonstrated nuclear fusion with the help of the proton synchrotron, a new type of particle accelerator that he had developed. The synchrotron was based on another brilliant idea from a few years earlier, Ernest Lawrence's cyclotron, but could accelerate particles to much higher energies. It's the same technique still used in the world's largest particle accelerator: the Large Hadron Collider at CERN. In the early 1930s, Oliphant was the first to bring about the successful fusion of two nuclei. He quickly realized that the resulting particles had much more energy than the particles that had collided, and that the new nucleus was lighter than the total weight of the fused small nuclei. These measurements, which Oliphant himself said were motivated by nothing but curiosity, suddenly made it clear that fusion of light nuclei could be what gives the sun its energy. Although the discovery of these fusion properties could easily have led humanity to its doom—since they form the basis for the biggest bomb ever made (the hydrogen bomb)—the same discovery may one day save humanity, because it has the potential to provide an almost limitless source of cheap, green energy.

The insight that energy is released when two light nuclei fuse was the last missing piece that we needed to figure out where the sun gets its energy. We already knew that the sun contained hydrogen and helium, and the temperatures inside it (millions of degrees) make hydrogen nuclei move so fast that they have enough energy to collide and fuse. It would take a while longer for us to fully understand the sun's energy cycle, but the fundamental process had become clear. All the hydrogen in the sun's core is gradually being transformed into helium nuclei, which transform in turn into even heavier nuclei such as lithium and carbon. This goes on until finally iron is produced. That's because the particles in the iron nucleus are so efficiently arranged that they take the least energy to hold together. An iron nucleus cannot gain energy by fusing or splitting. It's the nuclear equivalent of the ideal business model, and that makes it the final stop in the sun's fusion process. A large star stops burning as soon as all its fuel has been converted into iron.

Once we had figured out exactly which processes are involved, we could also calculate how and when our own sun's life would end. Sad but true: our sun will "go out" one day—in about five billion years, to be exact. But it'll be curtains for life here on earth at an earlier stage, when the outward pressure of the sun's radiation on all its particles becomes so great that the sun swells into a red giant and swallows up our planet.

Finally, just before World War II, the sun had given up its secrets. Besides showing us what fueled the sun, the discovery of nuclear fusion had other consequences. If the universe started out with vast quantities of hydrogen and helium, then the center of a star is the only place where other elements can be made, from carbon and oxygen to the very heaviest ones like iron. In fact, all carbon and oxygen atoms, the building blocks of all life on earth, were once made in a star's core.

Now that you know how the sun will "go out" one day, you might also wonder how it "came on" in the first place. Once the fusion reaction is in progress, it generates a lot of energy, but to get started, it requires temperatures of roughly 27 million degrees Fahrenheit (15 million degrees Celsius). Only then will atomic nuclei move fast enough to overcome the repulsive force between them, so that they can touch and possibly fuse. But how did the sun ever reach a high enough temperature for fusion to begin? Explanations in physics often seem magical, but fortunately, this one is simple. The sun formed because hydrogen gas drifting around in the universe started to clump together. Massive particles tend to do that, because they gravitationally attract each other. So they formed a large cloud of gas, which kept shrinking into a tighter and tighter ball. The growing pressure warmed up the gas in the center of the ball. As the pressure went on increasing, the temperature rose to millions of degrees, until it was finally high enough for fusion to begin. That was when the sun "came on."

One question I'm often asked is why the sun didn't burn up right away in one huge explosion as soon as it came on, the way gas does when you hold up a flame to it. For one thing, we're talking about a different kind of burning, because there's no oxygen in the sun. But it's still a good question. Ever since the first moment energy was produced inside the sun by nuclear fusion, the heat and radiation have been trying to move outward, pushing the gas along with them. But they can't, because so much gas is pushing inward at the same time and preventing the expansion.

But imagine if the radiation *were* strong enough to push the rest of the gas away. In this scenario, something interesting happens. Because the gas has moved outward, it's less compressed and the temperature is lower . . . so the atoms stop fusing . . . so nuclear fusion comes to a stop . . . so the gas is no longer pushed outward and collapses back inward . . . so the temperature increases again . . . so nuclear fusion starts up again . . . and so on. It's a subtle balance, which will keep the sun burning bit by bit—like a candle—until all its fuel is used up, 4.5 billion years from now. Although we don't necessarily need to know where the sun's light comes from to enjoy lying on the beach in France (or anywhere else), we do now know that "green" solar energy comes from nuclear reactions inside the sun.

There's one unique element of this story that you can use to impress your friends over drinks. As a star slowly burns up, the elements it produces get heavier and heavier, but it can't make anything heavier than iron. The only time heavier elements, like gold and silver, are ever made is during the death of a special type of star, which doesn't go out like a candle, but explodes. The next time you use a silver knife or wear gold jewelry, you might pause to consider the fact that every scrap of gold and silver was created when a star exploded. How did those metals ever end up here on earth? That journey took place over an incredibly long time frame, compared to which the entire history of human life on earth is a blink of the eye.

The energy hidden in the atomic nucleus is truly astonishing. If you managed to start up a controlled nuclear fusion process on earth, it would be a very efficient way of releasing a lot of energy in a short time. A civilian might see this as a promising solution to the energy problem, but in the military, they saw the potential for a big bomb. After World War II, there was still a lot of funding available for military research in the field of nuclear physics. But even though the hydrogen bomb had become a reality by 1952, we are still working on nuclear fusion as a source of energy now, more than sixty years later.

If you can bring two hydrogen nuclei together, you're in business. But that's incredibly difficult, because they have to be moving fast enough to overcome the repulsive electric force. The only way they can reach a high enough velocity is if the ambient temperature is around 15 million degrees Celsius, as it is in the core of the sun. It's hard enough to build a pizza oven, let alone an oven that can reach temperatures fifty thousand times greater. The truth is, the only way to achieve that on earth is by setting off an atomic bomb. That's exactly how a hydrogen bomb works. It's hard to believe, but a "regular" atomic bomb (a fission bomb) is used as a lighter, creating a temperature of ten million degrees for one brief moment. That makes the hydrogen start burning, releasing its energy in a chain reaction. In a fraction of a second, more energy is produced than that of a few hundred fission bombs. It was a formidable technical challenge, but the United States did manage to build a hydrogen bomb—thanks to Hungarian-American physicist Edward Teller's leadership and funding for atomic weapons even after World War II (inspired once again by the fear that "the other side" would get there first). On November 1, 1952, the first hydrogen bomb was detonated on the Pacific islet of Elugelab, part of the Enewetak Atoll. A few years later, the Russians succeeded in developing a similar bomb and setting it off. Since then, the two great nuclear powers have held each other in an iron grip. Five hundred to a thousand times as powerful as the atomic bomb used in Nagasaki, the hydrogen bomb clearly gave both sides the capability to wipe out humanity from that moment on. It's terrifying to realize how close we humans have sometimes come to the edge of the abyss. Take the Cuba crisis, when the United States and Soviet Union nearly started a nuclear war because of a conflict over the placement of nuclear weapons in Cuba by the Soviets. This is another consequence of the scientific secrets we found in that famous Pandora's box, the atomic nucleus.

Although the hydrogen bomb is undoubtedly the most destructive thing that humanity has ever developed, nuclear fusion has the potential to solve the world's current energy crisis. The great disadvantage of nuclear *fission* in power plants is that the materials consumed and produced are highly radioactive. Nuclear *fusion*, by contrast, starts with harmless hydrogen and ends with non-radioactive elements. If you could bring about a controlled fusion process, adding fuel in careful doses, then it could be an almost inexhaustible source of clean energy. The potential is enormous, but despite the large investments of money and research time (nuclear fusion receives more research funding than any other form of energy), we haven't yet been able to build a working, energy-producing fusion power plant. The remaining barriers to fusion power are mostly technological. The temperatures needed to start the fusion process and keep it going—millions of degrees—can be generated in plasmas. But because the particles in those plasmas have so much energy, the small fraction of particles hurled at the walls of the machine does considerable damage, making the walls radioactive.

To take a real step forward and show that fusion is possible in principle, the world's nations have decided to join forces and build one large facility in the south of France, the ITER project at Cadarache, the research and development center for energy research near the village of Saint-Paul-lès-Durance. Even though it will take at least ten years before ITER is operational, and probably a couple more decades after that before we have a working fusion power plant somewhere on earth, I have high hopes that we'll eventually be able to use the fundamental energy hidden in the nucleus in a constructive way.

Atomic and Nuclear Physics in a Nutshell

The atom is the smallest indivisible building block of the elements we find in nature, like oxygen, iron, and silver—it's a million times smaller than the smallest object you can see with the naked eye. It sounds like an impossible task to get to know that world, but with the help of particle accelerators, we have succeeded in mapping nature at that ultra-small scale. What we've discovered is that, at that level, nature has a logic of its own.

This fascinating journey, made possible by the development of particle accelerators, has yielded a wealth of new insights and applications. For example, we've discovered that everything on earth—in fact, everything in the universe—is made up of only three building blocks: protons, neutrons, and electrons. At the same time, the very peculiar phenomena we found in this new world completely transformed our picture of the laws of nature on the very smallest scale. From quantum mechanics to the energy of the particles in the atomic core, humanity has learned to harness these ideas. And as distant as they may seem from our everyday lives, they form the building blocks of both fundamental physics and our modern society, with applications in fields like computer technology and medicine.

These discoveries made scientists eager to learn more, and the new insights raised new questions. Are protons and neutrons really the ultimate, smallest building blocks of nature, or are there even smaller particles waiting to be found? And what lies behind the unsatisfying "logic" of quantum mechanics? Physicists wondered if they were missing something. And that wasn't all. Studies of nuclear forces had unexpectedly turned up all sorts of new, inexplicable particles that weren't protons, neutrons, or electrons. So what were they? In the decades that followed, new developments and discoveries took place in rapid succession, ultimately leading to the understanding of nature that we have today: the Standard Model, with three families of elementary building blocks even smaller than nucleons, and three quantum forces. In 2012, the long journey would finally culminate in the discovery of the cornerstone of the Standard Model: the Higgs boson. But at this stage, that was still a long way off.

3

Particles of the Standard Model

Ernest Rutherford was the first to find his way into the world of the atom, with its small nucleus orbited by electrons. And fairly soon after that discovery, we even uncovered a few of the secrets of the nucleus. We were making such rapid progress that it seemed only a matter of time until we could expose the world of the nucleus in greater detail with our ever-improving microscopes. We had learned about exploring and describing new territories from the European explorers of earlier centuries. When first traveling to areas like central and southern Africa, they had started with one-day expeditions from the safe havens on the coast before gradually penetrating farther and farther into the interior. Today, Africa has been crossed in every conceivable way, but we are still trying to get to the bottom of the particle world. And like the Europeans who found unfamiliar animals such as giraffes and elephants when they first stepped onto African soil, we have also been finding a variety of exotic creatures, at the subatomic level.

Scientists traveling to the very smallest scale, with an unknown destination, are very much like those explorers in

the olden days, journeying up a river that penetrates ever farther into a still-uncharted area. As they go deeper and deeper
into the jungle, they don't know how many hundreds of kilometers it stretches on ahead of them, or whether they'll find
anything but the same fruits, animals, mountains, and grassy
fields they have at home. The idea of being the first to set foot
on unknown territory is more than enough encouragement
for most people, but of course they secretly hope for more:
unique treasures and novel discoveries, like a golden city, new
species of animals, or lost civilizations. Fantasies like these, often inspired by adventure books like Jules Verne's, are the great
dreams that motivate explorers. And to keep the flame alive,
they hope for clues: a skeleton they can't identify, or a type
of riverboat they've never seen before, headed downstream
carrying golden slippers or drawings of a city deep in the jungle. That's when they know that more surprises lie ahead.

This is like the situation in particle physics right after
the discovery of nuclear forces. We had witnessed a variety of
strange phenomena for which we had no clear explanation—
clues that persuaded us to venture still deeper into the jungle.
But in that new world, nothing lasts longer than a billionth
of a second. So to study it at our own pace, we had to invent
new detection equipment and clever techniques. Eventually,
an amazing vista opened up before us. But it was clear we
still hadn't reached our journey's end, and in the search for
answers that followed, we gained an unprecedented wealth of
knowledge.

Until 1920, there was no sign of the surprises ahead. Physics seemed fairly neat and tidy. Neat and tidy *at last*, I should
say—because as we have seen, the changes between 1900 and
1920 had been nothing short of a revolution. We had fully
grasped the patterns in the elements and found out that all

the stable matter in the universe was made up of just three building blocks: protons, neutrons, and electrons. And we had uncovered a few of nature's most closely guarded secrets: the theory of relativity, quantum mechanics, and the nuclear forces. Even though we still didn't know the origins of that behavior, those laws of nature provided a firm foundation. With the help of a few simple rules, we could put together those three building blocks to assemble all the known elements, from oxygen to gold and from plutonium to mercury. First you take a few protons and neutrons and combine them to make a stable atomic nucleus. Then you send a few electrons into orbit around them, so that the whole thing is electrically neutral, and *voilà*: you've got yourself an atom. Neat, tidy, and surprisingly simple.

And yes, of course there were plenty of questions we couldn't answer. Where did those strange laws of quantum mechanics come from? Wasn't it strange that nearly identical combinations of those three building blocks gave us elements with such different properties? Take helium, a gas made of two protons and two neutrons. Add just one more proton and one neutron to the nucleus, and you get lithium, a metal. But even though many questions remained unanswered, it took the spectacular discoveries of the years that followed—the new phenomena and surprising particles—to make it clear to scientists that the end of their journey was still far ahead.

What we discovered is that, alongside the stable matter familiar to us, there's also a lot of matter with a lifespan less than a millionth of a second. A whole new world. This stage of our journey of discovery, when we stumbled across all those new particles, is what gave us the insight we would need to uncover all the particles and forces that we know of today. A second revolution!

In this section, it may seem as if I'm giving away the game before the chapter has even really begun. But my intention is to show you the big picture. This way, you'll be better able to appreciate the final step, and the joy that physicists felt when, in that chaos of hundreds of particles, we finally discovered the underlying pattern that brought order, calm, and simplicity to the world of elementary particles.

Discovering the new particles and working out the rules that describe their world was an adventure without equal in all of science: one involving surprises, disappointments, blood, sweat, and tears—followed by sheer euphoria. On this journey, our progress depended on two things: the new techniques and devices that we invented and built (like particle detectors and accelerators) and the power of the human mind to recognize patterns. To understand how the framework of elementary particles has expanded over time, it's helpful to keep in mind the following three steps.

> **Step 1: Nucleons (particles in the nucleus) are made of quarks.** The nucleons—the proton and the neutron—are not elementary particles, but are made up of even smaller pieces called quarks. In fact, you can also combine quarks into other kinds of particles. But the neutron and the proton are the only ones that last long enough to form the building blocks of our world's stable atoms. In spite of all our searching, we still haven't found any sign that the electron is made up of smaller parts. So the electron remains one of our elementary particles.
>
> **Step 2: Each family of particles includes one ghost particle.** We now know that alongside the three

building blocks of stable matter, there's another type of particle, the neutrino. It plays a key role in radioactivity, has almost no mass, and interacts with matter so little that it can fly straight through the earth without even noticing. Even though the differences between the neutrino and the electron seem huge, after years of investigation we discovered a common property that makes them inextricably linked. That's why we call both electrons and neutrinos *leptons*.

Step 3: Each elementary particle has a little brother and a little sister. There is a threefold symmetry. It turns out that each of the particles in the first family—the two quarks, the electron, and the neutrino—has two copies. These copies are heavier than the original particles, and highly unstable. Once they've been produced, the heavier particles quickly change into lighter forms, until eventually only the lightest and most stable of the three remains. We don't need the extra copies of the stable particles to understand all the materials we find here on earth. The stable particles are like regular potato chips. As long as your supermarket carries the traditional kind, you don't need the fancy salt and vinegar or barbecue flavors. Still, it's fun to know they're out there. Likewise, all those unstable particles give nature extra flavor. That threefold symmetry leads to interesting phenomena, but we still don't understand what causes it. Why do the copies of stable particles exist at all? And if there have to be copies, why not one, or ten?

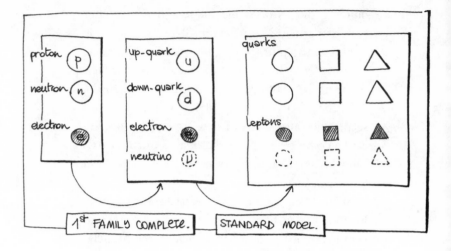

It comes down to this: There are twelve elementary particles, arranged in a neat system. They're divided into three families of four particles each: two quarks, a neutrino, and an electron-like particle.

At the same time that we discovered these particles, we also figured out the laws and characteristics that describe their behavior. We will now follow the journey that got us there. As you read, you can refer to the next illustration, which shows each step leading to the Standard Model, the final system of elementary particles as we understand it today.

Discovering New Particles, 1920 to 1970

The exacting work of mapping the subatomic world turned up all kinds of evidence that there are more things in heaven and earth than electrons and nucleons. We discovered that we are bombarded with radiation from outer space, called cosmic rays. We saw hints of strange, almost invisible ghost particles (neutrinos), and the electron turned out to have a heavier brother (the muon). Stranger still, we found out that we can *ourselves* make particles that exist for only a fraction of a second, and thanks to improvements in our measuring equipment, we learned more and more about the properties of these strange particles. And eventually, someone discovered the underlying pattern concealed in all those measurements: the Standard Model.

If we stuck to the exact historical order of events, it might get confusing—after all, the scientists themselves were pretty confused back in those days. So I'll jump back and forth in time a little bit. But first I'll give you an overview of the different ways that we particle physicists discovered new particles, using the analogy of a field biologist on an expedition. When a field biologist enters new territory in search for unknown animals he can discover new species in different ways.

1. **Direct observation** Suppose you're a naturalist exploring a newly discovered island. Before you head off on your expedition, you can't be sure you'll discover anything new there. Maybe the plants and animals will be just the same as where you are from. But once you're on the island and you find an animal with strange features (say, a unicorn with wings) you don't have to think long to realize that this is a new kind of animal you don't have at home. Still, before you can boast to your friends that you've discovered a new species, you'll have to prove that the animal has unique traits. Now, in the case of a winged unicorn, that's obviously not a difficult job. That's exactly what it was like when

physicists first spotted the muon particle. They reacted with the same surprise: "Huh? What the heck is that?"

2. **Looking at indirect evidence** As a field biologist, if you find tracks or hairs next to a watering hole that weren't left by any familiar species of animal, then you know there must be a new species somewhere around. At that stage, you haven't even seen it yet, but that doesn't matter. If it's a very timid species, it could take a long time before you actually spot one. But even before then, you can find out many things about the animal. By studying its tracks, you can learn approximately how heavy it is and whether it lives alone or in groups. And its droppings can tell you whether it's a carnivore or an herbivore. This is exactly the procedure by which particle physicists discovered the neutrino particle. The particle itself was invisible, but the indirect clues left no doubt of its existence.

3. **Fossils and remains** No one has ever seen a dinosaur, but thanks to their skeletons and other surviving fossils, we know they existed. If you go to a natural history museum, you can see those skeletons for yourself, and the many books on dinosaurs in the museum library will make it clear that we can figure out lots of things about the large, diverse range of dinosaur species. In the subatomic world, many of the inhabitants, the elementary particles, survive for only a billionth of a billionth of a second. That's much too short a time for us to see them,

let alone study them carefully. But fortunately, when they vanish, they turn into particles that *do* stick around long enough for our detectors to see them. So we can reconstruct that there must have been a heavy particle around, just as we reconstruct facts about dinosaurs from their bones. This technique has led to the discovery of hundreds of new particles—including the Higgs boson in 2012.

There are two essential tools for understanding this new world: the particle accelerator and the human brain. To understand how these two vital tools are used to study particles and their properties, let's return to our analogy of the field biologist, who is out to study and describe newly discovered animals in detail.

- **Breeding and studying animals in captivity**
  If you want to study a cow or a rabbit, that's no problem whatsoever. They have long lives, you can see them with the naked eye, and their behavior isn't especially complex. But consider the mayfly, which lives just one day in its adult form. Suppose you've run into this species for the first time. You haven't seen any live ones, only dead mayflies on the ground. And even if you catch a living mayfly, you won't have long before it dies. Mayflies also have delicate wings that make them incredibly hard to study. Imagine how easy it would be if you had a supply of immature mayflies in one corner of your lab, so that, as they matured, you could study many adult mayflies

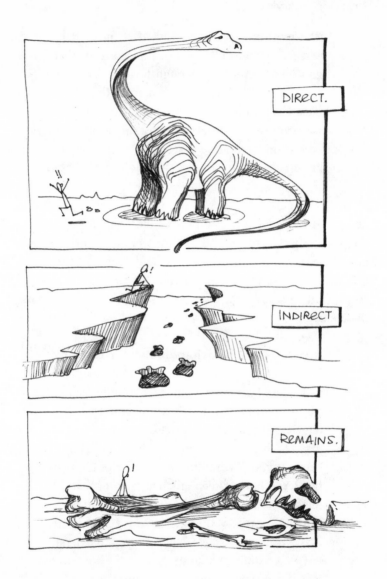

in a controlled environment. That's exactly what we've done in particle physics: we've created an enormous number of particles for ourselves.

The method of creating particles was one of the most important discoveries in this period. Albert Einstein's well-known formula $E = mc^2$ didn't mean only that you could turn mass into energy (like the energy created in the sun or a nuclear plant, discussed earlier). It also works in the opposite direction: if you bring together enough energy, you can make mass (particles). As our particle accelerators grew more powerful, we discovered these new possibilities. One advantage was that we could make all those particles in the controlled environment of our experiments. Although the particles themselves are too short-lived to observe directly, their remains do reach our detectors. That gives us enough information to determine all their properties.

- **Recognizing patterns and making predictions** Discovering new animals is great, but once you have a complete list of fauna, it's interesting to see whether you can find patterns. A first step might be to divide them into categories like fish, birds, and land animals. But then again, there are animals that live both in the water and on land, birds that can't fly, and so on. This tremendous diversity is what makes it possible for us to trace the origins of the variation we see. The most famous example is, of course, the small variations in species of finches that Charles Darwin ran into and studied on the Galapagos Islands.

This research gave him the puzzle pieces that
led to his great insight, the theory of evolution,
in which they all suddenly fell into place. Like
Darwin observing his finches, particle physicists
spent countless hours pondering the hundreds
of new, heavy, short-lived particles before they
arrived at their deep insight. Despite the great
diversity, they found that all those heavy particles were made up of only a few basic building
blocks: the quarks.

One direct result of seeing patterns (or
imagining that you do) is that you can make predictions. Once you discover a lioness with her
cubs, you know you'll find a male lion, even if
you've never seen one before. When we observed
how particles behaved in our experiments, we
noticed that, like most animals, they came in
two kinds: the particles of ordinary matter and
their antiparticles. So we made a prediction
about something that no one had ever observed
directly. A few years later, our prediction turned
out to be right: we discovered antimatter.

The second revolution in particle physics—from the atom to
the Standard Model of elementary particles—was possible
because we learned to "look" more carefully, and because we
discovered how to make new particles ourselves. The insights
that followed led to more than ten Nobel Prizes and changed
our perspective on the world forever. There are many new
terms to learn, but the most important thing to keep in mind
is that we succeeded in uniting all the particles, all the forces,
and all the phenomena we had observed in nature at the very

smallest level into one comprehensive framework: the Standard Model.

A story like this one, summing up how many different particles were discovered, can seem like a dry catalogue, especially if you're hearing it for the first time. But that's a misunderstanding. If you're the first European explorer, even if you've already discovered a giraffe and an elephant, you'll still be impressed when you find your first crocodile or penguin. Every new species, and every new type of particle, presents a unique experience and adds to our knowledge. So the puzzle was greater than we thought after we had cracked the atom. What are all those things we found, and what do they tell us about the foundations of physics? In short, what lay deeper in the jungle?

Let's start by discussing some of what we have observed *indirectly*. Although I have already described the phenomenon of radioactivity in detail, and we had an ever-improving understanding of alpha, beta, and gamma radiation, there were still great uncertainties about the exact underlying mechanisms. One of the greatest mysteries was the energy of the radiation in beta decay, in which one of the neutrons in the nucleus spontaneously transforms into a proton and an electron. The proton stays put in the nucleus, but the electron is sent flying off at high velocity and forms the radioactive beta radiation that we feel. Although the mystery mainly has to do with the electron, it's worth noting that when the nucleus goes through this transition, the atom's identity changes. After all, the number of protons increases by one, and as we know, that number determines which element the atom is.

But anyway, we were talking about that runaway electron. Using well-established principles of physics, you can determine fairly easily the energy with which the electron will fly off. It's always the same, and since we know the mass of both the electron and the proton, we can calculate that energy with great precision. So imagine our surprise when we discovered that the electron's energy was not what we had predicted. In fact, the energy of the electron was different each time beta decay took place. The laws of nature, as we understood them back then, made this result impossible. And because the measurement technique used in the experiment was fairly simple, it was hard to imagine where the mistake could lie. Those measurements were not going anywhere. It was up to the theorists to find a solution.

Finally, in 1930, Wolfgang Pauli stuck his neck out. He showed mathematically that it was fairly easy to explain the range of electron energies, but only if you assumed that when the neutron fell apart, not only the proton and electron were released, but also a third particle. The energy and momentum would then be divided over three particles rather than two. Since the electron and the new particle *together* had to cancel out the movement of the proton to preserve the balance, the electron had more or less energy depending on the direction in which that new hypothetical particle flew off. A masterstroke—Pauli's mathematics described the exact distribution of energy of the electrons observed in experiments. And because the measurements were so clear, it was possible to go right ahead and determine the properties of that new particle.

But there was one little hitch. The apparent properties of the new particle were nothing—and I mean *nothing*—like the properties of all known particles. The fact that it had almost no mass and no electrical charge was a bit peculiar, but posed no real problem. What *was* a problem, though, was that no particle like it had ever been observed in any experiment on earth. Unusual, to say the least. After all, the only possible explanation would be that this type of particle, assuming it exists, barely interacts with the matter we know—so it doesn't even notice we're there, so to speak, and flies straight through our detectors unobserved. That last property—flying through walls and steel without being stopped—is what makes it a real ghost particle. Pauli's theory felt like a drastic solution and made him, and his fellow physicists, uncomfortable. It's like a detective solving a murder mystery by eliminating possibilities until the only remaining hypothesis is that a ghost stole the jewels by flying straight through the wall into the room and opening the safe.

As a true ghost particle, the particle's existence would be almost impossible to confirm experimentally. Was it really a particle at all, or just a mathematical trick, and wasn't there some other, more plausible explanation? Pauli didn't even want to publish the idea, and when he presented it to other scientists, his tone was apologetic. Essentially he said, "Ladies and gentlemen, I have done something terrible today. I have come up with a desperate remedy by proposing that a new particle exists that cannot be measured. That's a thing that no theoretical physicist should ever do." Fortunately for Pauli, his colleague Enrico Fermi did take the idea seriously and developed a theory of beta decay in which Pauli's neutrinos (Fermi came up with the name) played a central role.

It must have come as a huge relief to Pauli when, in 1956, two American scientists, Frederick Reines and Clyde Cowan, finally proved the existence of his undetectable ghost particle. Finally, a direct measurement instead of indirect measurements and clues, which had always seemed a little unsatisfying. But how did they pull that off? If this particle hardly interacts with matter, then how can you see it? The conventional solution would be to build a detector, set up the experiment, and then wait patiently for a hundred million years in hope of witnessing a single neutrino collision. So, not an option. Luckily, physicists are a surprisingly inventive breed, and they came up with a trick. If any one neutrino has very little chance of colliding with your measurement equipment and leaving a trace, then you have to make sure to fire *lots and lots and lots* of neutrinos at the equipment so that you'll eventually see a couple of traces.

If you were to take Fermi's new theory of beta radiation and nuclear interactions seriously, then what you would need to do is produce an enormous quantity of neutrinos in a nuclear reactor, one for each unstable atomic nucleus that decays. By building a detector close to the reactor, you should be able to see a very occasional reaction when one of those many neutrinos happens to hit a nucleus. We won't get into exactly what happens then. The important thing is that the collision should produce a very characteristic flash of light that you can easily measure with an ordinary camera. Reines and Cowan set up an experiment close to a nuclear reactor and actually observed the flash predicted by the theory. Neutrinos existed! Pauli had already received a Nobel Prize, in 1945, for his famous exclusion principle (which explained the numbers of electrons in the orbits around the nucleus), but Reines received the Nobel Prize in 1995 for the direct experimental discovery of the neutrino.

At this stage, it's important to mention two things. The first is that there are not one, but three different kinds of neutrinos in nature. The neutrino we've been discussing so far is known as the electron neutrino, because it turned out to be closely related to the electron. Later, we'll get to the detailed story of how two other particles were discovered that are a lot like electrons: the muon particle and the tau particle. And just like electrons, these particles were found to have neutrinos as partners. You won't be surprised to hear that we call them the muon neutrino and the tau neutrino.

Research into and with the three neutrinos has been a hot topic in physics in recent decades, and that trend will continue in the years ahead. There's a lot we still don't understand, and looking at neutrinos could lead to some interesting answers. For example, we know for a fact that neutrinos have a tiny mass, but even though we realize they must be a million times lighter than electrons, their exact mass remains a complete mystery. And even if we could measure their mass, we still wouldn't know *why* that mass is a million times smaller than an electron's. Is it "just how things are," or is there some system behind it?

On top of that, neutrinos are the David Bowie of particle physics. They can keep transforming between the three different kinds (this is called oscillation) and can even be in a mixed state. The existence of these neutrino oscillations was proved by Takaaki Kajita and Arthur B. McDonald, who won the Nobel Prize for this achievement in 2015. In the United States, one of the world's largest neutrino detectors is being constructed for the Deep Underground Neutrino Experiment (DUNE). The goal is to see whether the mixed state and the properties of neutrinos help to account for the lack of antimatter in the universe. There are also various experiments aimed at finding the exact mass of those ultralight neutrinos.

Because neutrinos are the only particles that can reach the earth unharmed from distant celestial bodies (photons are stopped by gases because they collide with the electrons in the atoms drifting around there, and protons are deflected by the magnetic fields in the universe), they also make an ideal telescope for observing celestial objects. For that purpose, we need detectors that are truly immense. Right now, there's one huge detector in the ice at the South Pole, and a large group of scientists is working on placing another detector, one cubic kilometer in size, on the floor of the Mediterranean Sea.

Observing and Identifying Particles in Experiments

Although our human senses are very well adapted to observing the world around us, they have their shortcomings. Both their range and their precision are somewhat limited. For example, we know that dogs can hear sounds we're unaware of. That explains the strange phenomenon that dogs will sometimes run off together or start howling for no apparent reason, all at once. They're hearing a sound that's hidden from human ears. Dogs can observe other things we can't. Don't bother asking a human to sniff a suitcase at the airport for traces of cocaine, but a dog will pick that suitcase out with ease. And if I ask you to give me the temperature to a tenth of a degree in the spot where you're reading this book, the best you can do is guess. But thanks to our ingenuity, we humans have overcome some of our shortcomings, by building machines that can detect all the things I just described. It's a straightforward idea, but often a difficult job—after all, how can you show that something really exists if you can't see it?

Give it a try sometime. Next time you're with your friends in a bar or at a birthday party, say to them, "I think there are invisible rays in the air here that transmit sounds and pictures that send messages." Don't be surprised if you get some funny looks. But you're absolutely right. There really are electromagnetic waves moving through the air all around us, which carry radio, television, telephone, and Wi-Fi signals. Your hands, eyes, and ears won't help you to prove that, so you'll have to find some other way. First of all, you'll need one device that can "sense" the rays (an iron rod, in this case—an antenna that picks up the waves) and another that can convert the signals into something we human beings can see or hear (like a radio, television, smartphone, or tablet).

Forty years ago Peter Higgs—and, at the same time, a pair of Belgian physicists—made a strange prediction. Like a true gentleman, Higgs didn't do that in a bar but in a prestigious scientific journal. He claimed that empty space isn't empty at all; instead, there's a field throughout the universe (we'll come back to what a field is later; for now, see it as a kind of thin, elusive substance) that gives particles their mass. This prediction, which seemed just as eccentric at first as claiming that there are rays all around us that contain voices and images, turned out to be harder to prove than the existence of radio waves. It would take more than forty years before we succeeded in demonstrating the existence of this field, the Higgs field, experimentally.

While an iron rod and a radio are all we need to detect electromagnetic waves, we had to develop a whole set of new techniques to discover the Higgs field. First we had to invent the most powerful particle accelerator ever, the Large Hadron Collider at CERN in Geneva, which has a circumference of more than twenty-seven kilometers. Next to it, we needed a detector, a camera the size of the White House in Washington, D.C., to study the fragments left behind when particles collide. The only way to make this scientific quest possible was for thousands of scientists from all over the world to join forces and give it a try together. And that's what they did. To achieve their goal, they would have to not only work cooperatively but also make the Higgs particle themselves and overcome problems of many different kinds. These were problems no one knew how to solve when the adventure began, problems that would require new tools to find a way forward. Before we go on to the story of this quest, and how we finally discovered that empty space is not really empty but filled with the Higgs field, it's important for you to have a solid understanding of the tools we use.

THE PARTICLE DETECTOR

The mission of particle physicists is to make elementary particles and their properties visible. These particles can't normally be seen with the naked eye, so we have to come up with other ways of making that small world visible and, if possible, manipulating it. In Chapter 1, we saw that you can learn things about small objects by firing even smaller projectiles at them and watching how those projectiles bounce off them. We also learned that you can make the projectiles smaller by making them move faster. In fact, the main purpose of a particle accelerator is to do just that.

When Rutherford fired alpha particles at a thin sheet of gold foil to study how an atom looked, it was essential for him to record the angles at which the particles bounced back. Since those particles were too small to see, Rutherford used a plate coated with a substance that causes a flash when hit by a charged particle. Everyone over the age of thirty remembers how televisions used to work: a beam of electrons was fired at a screen consisting of very small pixels, each of which was made up of three dots—red, green, and blue—that lit up when hit. Computer screens still work with those same three basic colors (RGB). So Rutherford's assistants sat in a dark room and recorded the angles at which the alpha particles were deflected. This came down to simply counting, and the distribution of angles revealed the structure of the atom.

But in Rutherford's experiment, how do you know how fast the alpha particle was moving and exactly which way it went? And how do you see the difference between a proton and an electron? What we need is a machine that not only shows in what direction a particle moved, but also tells us as much as possible about its properties, like velocity, mass, and electrical charge. When I give talks, I often use the analogy of footprints in the snow. If I show people a trail of footprints in the snow and ask whether they were made by a car, a rabbit, or a human being, they think it's a strange question: "A human being, of course." And a closer look at the photograph can tell us even more. We're capable of figuring out whether it was one person or two, whether it was a child or an adult, and many other facts.

In a particle detector, we do much the same thing. Whenever a particle passes through the detector, it leaves behind a telltale pattern, like those footsteps in the snow. That pattern reveals one or more of the particle's properties. By sending particles through several detection layers, each of which is designed to identify a particular property, you can collect different pieces of information about the same particle. And by combining that information, you can form a clear picture of each particle that flies through the detector. Just as you can reconstruct dinosaur skeletons from bones to learn more about their world, you can reconstruct a portrait of heavy particles from the fragments you find in your detector.

THREE ESSENTIAL ELEMENTS OF
A PARTICLE DETECTOR

Basically, a particle detector has three components for three kinds of measurements.

1. **Make the path of electrically charged particles visible.** When an electrically charged particle moves through a material, the only part of the material it notices is the electron clouds. Those slow down the particle slightly, making it lose a little bit of energy. That loss of energy is due to various processes, but at low energies, the main one is ionization: the particle strips off an electron from one of the atoms that it passes. The atom in question then has a positive charge (in other words, it's ionized) and the electron floats freely. You can take advantage of this phenomenon to make the particle's path through the material visible, as we'll see.

   Around the time of Rutherford's discovery, the Scottish physicist Charles Wilson invented what we call the cloud chamber. He filled a closed container with water vapor in a very specific state: supersaturated and supercooled. That means that there was more fluid in the air than it can normally hold, so that the least disturbance was enough to change the vapor into a liquid: condensation. A charged particle that ionizes atoms on its way through the chamber is exactly the kind of disruption that sets off that process. The ionized atoms form what are known as condensation centers: the points of disruption where the vapor turns into a fluid and a small cloud of droplets appears. It's a little bit like the phenomenon of a condensation trail high in the sky on a clear day, which shows you that an airplane is passing even if the plane is too far away to see. And this thin trail of clouds, which forms only at great heights and under special conditions, often remains for a long while after the plane is gone. Wilson's cloud chamber was a similar way of making particles visible (years before airplanes could reach those heights). This technique made the charged particles we'd already discovered—like protons, alpha particles, and electrons—visible to the eye, so that you could observe and analyze them. The cloud chamber played an astonishingly important role in particle physics and won Wilson the Nobel Prize in 1927.

A cloud chamber is not really high tech. Any high school can build one, and there's one in the lobby of my workplace, the National Institute for Subatomic Physics (Nikhef) in Amsterdam. Since there's always natural radioactivity around us, you can always see alpha particles and electrons moving through the chamber, but if you want a more spectacular show, you can place a radioactive source near the chamber. It's fascinating to see radioactivity so directly, with the naked eye. In the 1950s, the cloud chamber was replaced with an improved version, the bubble chamber invented in 1952 by Donald Glaser.

Because it works with hot liquid instead of vapor, a trail of bubbles appears—like the bubbles in a pot of boiling water—instead of condensing liquid, but that's just a detail. The bubble chamber is more stable and creates a sharper image, but the final result is still a photograph of tracks (in this case, bubbles) created by the passing particles. Glaser won the Nobel Prize for his invention in 1960.

In modern experiments, we don't generally use gas or fluid. Instead, we send particles through very thin plates of silicon. When electrically charged particles pass through them, they create free electrical charge in the material. If you generate a strong electrical field in the material, then the electron and the ionized atoms, moving in the opposite direction, will cause a small current that you can measure. If you make those sensors small enough and stack them up, you can "connect the dots" and very easily and accurately see what path the particle followed. In the illustration a little later in this chapter, the particles fly from left to right, and the detector layers are indicated. The traces left by a particle in all these layers are marked with Xs. Once you have that data, it's not hard to interpret it as the track left by a particle flying past. That path is also shown in the picture.

One important part of this story is that the amount of energy lost by a particle (the number of bubbles or the strength of the current in the silicon) depends not only on a particle's energy, but also on its type. It was this property that led to the first discovery: a particle was seen that lost less energy than a proton or an electron. In other words, a completely new particle.

2. **Determine the charge and momentum (velocity) of electrically charged particles.** Electrically charged particles are deflected in a magnetic field. The force acting on them is called Lorentz force, named after the famous Leiden physicist Hendrik Lorentz. He showed that in an area with a magnetic field, a charged particle moves in a circle. The radius of the circle depends on the particle's velocity. The faster the particle moves, the harder it is for the magnetic field to deflect it, and the larger the circle will be. This is just like cars driving around a traffic circle: the faster the cars are traveling, the larger the circle has to be to keep them from going off the road. If you know how strong the magnetic field is, you can determine how much the particle is deflected—in other words, the radius of the circle. Then you can calculate its velocity.

In the sketch of particles flying through a detector, you can clearly see the difference between a slow particle and a fast one. Another convenient fact is that particles with a negative charge, like electrons, are deflected in exactly the opposite direction from positively charged particles like protons.

3. **Measure the energy of particles by forcing them to a stop (calorimetry).** Since particles lose energy in a thin layer of material, a very

thick layer can bring them to a complete stop. For a good analogy, think of throwing an iron ball into a large block of Styrofoam. If you toss the ball into the foam casually, it will go only a few centimeters deep, but if you throw it with tremendous force, it could make a hole as much as a meter deep. In other words, how deep the ball penetrates says something about the velocity at which it flew into the material. A similar phenomenon is used in particle detectors. There too, we bring the particles to a complete stop in a heavy substance. A certain fraction of the energy lost as the particle slows down takes the form of light. By measuring exactly how much light is produced and how deep the particle went, we can estimate the energy of the particle as it hit the block.

One important point is that to stop all the particles, you need two separate "energy meters": an electromagnetic calorimeter and a hadron calorimeter. The best material for bringing electromagnetic particles (electrons and photons) to a halt is almost transparent to proton-like particles (hadrons), which fly through it quite easily. To stop them, you need a thick layer of extremely heavy material. This type of detector is called a hadron calorimeter.

I should add that measuring the light produced as the particles slow down is no easy job. The problem is that the light is produced in the middle of a block of heavy material. How do you get it out of there, so that you can measure it with a camera? In an ideal world, you use a material that is both very heavy (that is, it contains many electrons and the particle is slowed down fast) and transparent (so that you can clearly see the light produced as it slows down). Strangely, but fortunately for us, there *are* materials that combine these two properties. Leaded glass, for example, with occasional atoms of lead incorporated invisibly into the crystal lattice of the glass. It looks just like an ordinary block of glass, but if you try to pick it up, you realize it's incredibly heavy. It's a unique experience the first time you feel its weight. We can capture all the light produced by the block as it slows down the particle by attaching a light detector to the end of the block.

The second type of calorimeter, for slowing down heavier particles, uses even heavier material, often iron or something similar. That's not transparent, so we have to be more creative. We use light meters called scintillators, built into the material, one every centimeter or so. A scintillator is like a thin layer of plastic that can capture light and send it, through an optical fiber, out of the block to a light detector (photodetector). So the entire hadron calorimeter looks something like the inside of a many-layered cake: layers of metal alternate with scintillators that collect the light and lead it away.

By combining the information from the different detectors, we can figure out important properties like electrical charge,

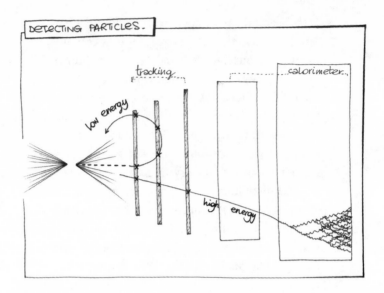

direction of movement, and energy. That tells us the type of particle. These detectors have enabled us to dive deeper into the world of the very smallest things, mainly by revealing phenomena that would otherwise remain invisible to us. Over the years, the techniques have improved, but the basic components are still included in every single detector.

Equipped with our ideas and our detectors, we can now go on with our adventure. What were the first strange observations we made, and how did those clues lead us toward a deeper layer of elementary particles?

Discovering New Particles through Direct Observations and Remains

In the late nineteenth century, one peculiar phenomenon was irritating physicists like a pebble in a shoe: the spontaneous discharge of electroscopes. An electroscope is probably the simplest piece of scientific equipment there is. It consists of two paper-thin metal plates hanging from the end of an iron rod in a vacuum inside a bell jar. The rod sticks out of the top of the jar. You can give the metal plates an electrical charge. This works the same way as the static electricity in that old trick where you rub a balloon against your clothes and it sticks to things or makes your hair stand on end. As soon as the plates in an electroscope are charged, they repel each other, because the two plates have opposite electrical charges. An electroscope is a fun way for high school classes to learn something about electricity. One well-known phenomenon was that the plates would gradually lose their electrical charge if radioactive material was close by. The radiation ionizes the air, inducing a small leakage current and thereby dissipating the charge from the plates. So after a while, the plates would move back toward each other. But why would the same thing happen even when there were no nearby radioactive sources?

Many thought that a small underground source of radioactivity was making electroscopes lose their charge. That sounds plausible. Case closed, you might think—but if you know any physicists, you've probably seen how stubborn they are. One of them always asks, "Is that really true?" and then goes off in search of a detailed answer. Theodor Wulf, a German priest who taught in Limburg in the early twentieth century, was just that sort of physicist. He built his own electroscope and started measuring. If the radiation really was coming from radioactive elements in the earth's crust, then an electroscope at the top of the Eiffel Tower would lose its charge more slowly than one in the underground corridors of the Limburg limestone quarries. Wulf went down into the quarries right away and, yes, also took a trip to Paris. But to his astonishment, the charge was lost almost as quickly at the top of the Eiffel Tower as at the bottom. That implied that a different, unknown, source of radiation was causing the discharge.

The height of the Eiffel Tower is impressive, but of course, you can go even higher. At the same time that Rutherford was experimenting with atoms, the German physicist Victor Hess also built an electroscope, took it up in a hot air balloon, and made lots of measurements at different heights. He saw that the discharge slowed down somewhat as he rose higher and higher, but to his surprise, after he passed a height of about five kilometers, the discharge started speeding up instead. The higher the balloon, the more radiation. It completely contradicted the hypothesis that radioactivity comes out of the earth. Only one conclusion was possible: the radiation comes from outer space. Hess called it *Höhenstrahlung* (high-altitude radiation), and even though he had no idea what it was or where it came from, exactly, its extraterrestrial origins were obvious.

We now know that the earth is constantly bombarded by protons produced somewhere in the universe. They collide with oxygen and nitrogen atoms in the upper atmosphere, producing electrically charged particles that rain down on us from the sky and discharge the plates in our electroscopes. This radiation, now called cosmic rays, is still a major topic of research, which tends to focus on the type with the very highest energy. Where do these cosmic rays come from, and what process sped them up to such high energies? Hess received the Nobel Prize in 1936 for his discovery, sharing the honor with Carl Anderson, the man who had opened the door to a new world. Anderson had discovered, in 1936, that there were particles hidden in cosmic rays unlike any we had ever seen before. These new particles would form the key to the discovery of the Higgs boson some seventy-five years later. Their name: muons.

Research into cosmic rays attracted a great deal of interest. What were those strange particles from space that kept hitting our world, and where did they come from? We had already developed instruments for studying protons, electrons, and photons, using our knowledge that each of those particles behaves differently in a cloud chamber. When scientists looked at which particles passed through the cloud chamber, they saw electrons, which were deflected in a magnetic field as they had predicted, as well as alpha particles from radioactivity, protons, and photons (particles of light). But—and here comes the big surprise—in 1936 they also saw a different particle: a particle that was deflected in a magnetic field, like an electron (and therefore had to be electrically charged), but behaved in a fundamentally different way. It didn't curve as sharply as an electron, left different tracks, and unlike electrons could easily fly straight through a sheet of iron. Since it also behaved differently from the only other known charged particle, the proton, it was clearly a new particle with unique properties. It turned out to be about two hundred times as heavy as an electron and was given its own name: the muon.

The muon, also called the electron's heavier brother because they're so similar to each other, was a particle that had literally come out of the blue. The muon was a new addition to the palette of elementary particles, which was gradually growing. Although electrons and muons share many properties, the most conspicuous difference is that the muon is about two hundred times as heavy as an electron and doesn't lose a lot of energy when it passes through matter. Another difference is that unlike the electron, which seems to live forever, the muon has an average lifespan of about 2.2 microseconds. After a little more than a millionth of a second, it falls apart into one electron and two neutrinos.

In the mid-1970s, researchers in California discovered that the electron and the muon had a little sister, which they dubbed the *tau* particle. Or maybe "little" isn't the right word. The tau weighs about seventeen times as much as the muon, and its lifespan is about a million times shorter. Like the electron, the muon and tau particles were each paired with a neutrino. The tau neutrino was discovered only very recently, in the year 2000. It had been the last missing particle of matter in the Standard Model.

The muon would play an important role in the discovery of the Higgs boson. Of all the new particles created when particles collide, muons can be detected most easily and accurately. And heavy particles often produce muons as they decay.

Although the cosmic rays would later lead to new insights, physicists were most interested in learning more about the nuclear forces and building a description of quantum mechanics. They were still puzzled by the remarkable fact that electrically charged particles are packed together in the nucleus even though the electromagnetic force should push them apart. Apparently there was another force at work that was stronger than electromagnetism. At the same time, we knew that this force did not operate outside the nucleus, because otherwise all the nuclei in the world would form one big clump.

In 1935, Hideki Yukawa found a solution to this puzzle that built on the description of the electromagnetic force. At the quantum level, electrical force is described in terms of exchanges of a messenger particle, the photon (which is the particle of light). According to this description, the heavier the messenger particle is, the smaller the range of its force. In the case of electromagnetism, the photon has no mass, which means that in principle the range of its force is infinite. If you want a new quantum force with a limited range, you need a messenger particle (in other words, a force carrier) that can't go beyond the nucleus. There could easily be such a messenger particle, as long as it's heavy enough. It was simple enough to predict the properties of this hypothetical particle, named the *pion*: it had to be about one hundred times as heavy as an electron and slow down fast when it encountered an atom.

When the muon was discovered, scientists were excited. Could this be the pion they'd been waiting for? But their hopes were dashed: it turned out to weigh not one hundred but two hundred times as much as an electron, and it didn't slow down in matter, but flew through it with relative ease. Argh! How could this be?

In the end, the pion was discovered thanks to another stubborn physicist. Following the lead of the discoverers of cosmic rays, Cecil Powell decided to go high in the mountains, in 1947, to measure how many muons were there. When he looked at the results of his experiment, he found signs of not only muons but also another type of particle: a particle that not only proved to be the source of the muons created there in the upper atmosphere, but also had exactly the properties that physicists were looking for. He had found the pion. The man who had made the prediction, Yukawa, and the experimental physicist Powell were each awarded the Nobel Prize, in 1949 and 1950, respectively.

Pions are created in the upper atmosphere when cosmic rays hit atoms, but they never reach the earth's surface because they come to a standstill way before that. When pions decay, they produce muons that *do* reach the ground. The prediction, inspired by developments in the quantum mechanical description of another force, combined with the actual observation of the particle high in the mountains, make this match a draw between theory and experiment. Experiment was soon to score a major victory, but after that, theory would make its comeback.

We've now reached a point where the number and variety of particles might make you dizzy, unless you keep in mind the fairly compact model that is our final destination. After all these surprising discoveries, the most important insight was that there were more types of particles than we had first thought. We had found the muon particle (the electron's little brother), the neutrino or "ghost particle," and the peculiar pion particle. We had also realized that some particles, such as atoms, are not elementary but can fall apart. We had seen that a pion produces a muon when it decays and that, at the end of its short life (2.2 millionths of a second on average), a muon produces an electron. One major implication of these findings is that by reversing the process you could make particles, through particle collisions. The particle accelerators available then were not strong enough to accomplish that feat, but fortunately, it proved possible to use high-energy cosmic rays as projectiles. And sure enough, when those rays hit thin plates of iron, new particles were produced. This marked the dawn of an era in which many strange new particles would be discovered.

In this period, physicists managed to build increasingly powerful particle accelerators, which freed them from relying on the whims of nature. It had been frustrating for scientists, having to wait and see whether nature would happen to send a cosmic ray through their detector, without even knowing its energy level beforehand. Now they could set up their own particle collisions in a controlled environment. Like the mayflies bred in a biologist's lab, new particles could now be systematically produced with certain properties and studied in detail. As we'll see in the rest of this chapter, the process of unraveling all those collisions and classifying the new particles led to a flood of new information. It seemed overwhelming at first, but we eventually saw that matter, despite its great diversity, is ultimately made up of just a few kinds of particles, and that the three forces of nature controlling their behavior at this distance scale follow a reasonably simple set of rules: the Standard Model.

Two final steps led us to that destination: the discovery of a tidal wave of heavy, very short-lived particles produced in collisions, and the realization that nucleons (protons and neutrons) are made of quarks.

At this point in our story, because of the destructive power of the nuclear bomb, nuclear physics had become a major field of research with huge funds at its disposal.

The discovery of the pion in 1946 had been a great step. If new particles could be created by collisions between cosmic rays and atoms in the upper atmosphere, then maybe we could study those collisions in our labs—for example, by exposing a plate to cosmic rays close to or even inside our detectors. The plan worked: inside the detectors, as we had hoped, we saw new particles appear under the plate hit by the cosmic rays. When physicists took a close look at these new particles, it sometimes seemed as if some pairs of particles (one positively and one negatively charged) appeared out of nothingness at a single point, about one centimeter *beyond* the plate itself. But

particles appearing out of nowhere—of course that was unthinkable. One proposed explanation was that a neutral particle had been made in the collision. Being electrically neutral, it
would give off no signals and remain undetectable until it split
into two charged particles. Since those particles were charged,
they were, of course, visible to our detectors. That made it look
as if they had come out of nowhere. From the characteristics
of the charged particles, we could infer many properties of the
neutral mother particle, like its energy, mass, lifespan, and so
forth. We then found out it was not the only new particle produced in collisions.

Even today, when we want to prove the existence of new particles, we still look at their decay products. Let's return to our
example of bombarding an iron plate with cosmic rays. You
could clearly "see" that the two new particles appeared to come
out of nowhere. That was caused by the neutral particle, which
survived long enough to travel one centimeter, even though it
was moving at about the speed of light.

But most new particles turned out to have much shorter
lifetimes, often as much as a billion times shorter. In a case
like this, even though you can't directly see the mother particle
moving, you can still conclude from the decay products that a
heavy particle was created in the collision. This is the standard
technique for identifying short-lived particles, and it's how we
later proved the existence of the Higgs boson. So it's important for me to describe each step of the technique in detail and
show you how we arrived at that conclusion. A professional
particle physics analysis, condensed into just one page.

One of a particle's most important properties is its mass.
When a particle falls apart—or, more precisely, *decays into
other particles*—its mass is preserved. Even if the daughter
particles have much smaller masses, you can use their ener-

gies and trajectories to calculate the mass of the mother particle. If you calculate this *invariant mass* for every collision with daughter particles of this kind, you should see the same value each time: namely, the mass of the mother particle. If only it were that simple. But every collision produces more than just the mother particle's decay products. That gives you an enormous number of possible combinations, only one of which is right. To make things worse, there are also collisions in which a particular type of mother particle is *not* created, but two particles of the same type as the daughter particles *do* happen to be produced. But in a case like that, the two particles have nothing to do with each other, so if you measure the invariant mass, the result will be an arbitrary value: sometimes low, sometimes high.

To show that a new heavy particle was lurking in a few of those many cases that you observed, you have to do two things after each collision: (1) hunt for the expected decay products, that is, the two daughter particles, and (2) calculate and graph their invariant mass. In collisions that don't involve a new particle, the masses will show a fairly flat distribution—sometimes high and sometimes low. But if a heavy particle is created, the value will always be the same.

The next figure includes a graph on the right, which shows the invariant mass for a large number of collisions. The horizontal axis represents the measured mass, and on the vertical axis is the number of collisions observed with that mass. The horizontal rectangle shaded in at the bottom shows the situation if no new particle was created: a flat, arbitrary distribution. But whenever the collision produced a new particle, the measured mass should be equal every time—*approximately* equal, that is, since no experiment is perfect. So you expect to see many more events in that range of the graph, caused by the newly generated particles. Until you run the experiment, the

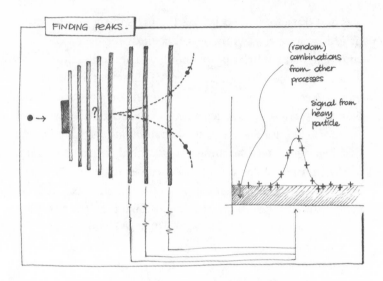

existence of the new particle is only a hypothesis. In this particular experiment, we see that the measurements (the Xs on the graph) don't show a flat distribution at all; instead, there's an obvious peak in the graph. A new particle.

The peak's horizontal position tells you the mass of the new particle, and its height tells you how often, in all those collisions, a new particle was created. The graph sketched here is exactly the kind we used in searching for all kinds of new particles, and in looking for evidence of the existence of the Higgs boson. That may all seem straightforward enough, but in my example, I exaggerated the strength of the new signal to make the principle clear. Once in a great while, we end up with a graph like that, but fortunately nature often makes things a little more challenging for us than this sketch suggests. Then we simply have to work harder to solve the puzzle.

We use a lot of statistics to figure out whether the peaks we think we see are real, or whether the deviation could have some other cause. For example, if we see a few more collisions in one spot on the graph, we try to figure out whether it's a coincidence, a stupid mistake in our calculations, or a glitch in the detector. Before we decide we've proved the existence of a new particle and tell the world what we've accomplished, we have to follow a very strict protocol. The burden of proof is so great because the loss of credibility would do so much damage. Some scientific fields have come under fire in recent years when claims, announced with great fanfare, could not be reproduced.

Using Particle Accelerators to Make New Particles

If there's one thing that annoys physicists, it's not being able to control and manipulate every facet of an experiment. Understanding nature is hard enough without all sorts of irrelevant phenomena getting in the way. But back in the 1950s, physicists had no choice. Their most powerful particle accelerators, the cyclotrons, had helped us to split the atom—and even today, they do a decent job of generating X-rays and radioactivity in hospitals. But even though they were increasingly powerful, they were not powerful enough to reach the high energies needed to reproduce the effect of cosmic rays, creating new particles in the lab. The only way to achieve collisions that *were* powerful enough was by using actual cosmic rays. The downside was that we didn't know exactly when the cosmic rays would reach our experimental equipment, what kinds of particles would hit the plate, or how much energy they would have. The solution came in the late 1950s, when we first achieved our own high-energy particle collisions.

It was the *synchrotron*, invented by Mark Oliphant, that allowed us to take the next step in particle energies. Here's Oliphant's idea: Particles are deflected in a magnetic field, so if you want particles to move in circles in a small, hollow vacuum tube, you have to make sure the magnetic field is exactly strong enough to send the particles on one complete circuit around the tube, ending right where they started. Accelerate the particles too much in some part of the tube, and they'll fly off the curved path. The magnetic field can cope with slightly slower particles, but it's too weak to deflect the faster ones. But suppose that as you accelerate the particles, you also strengthen the magnetic field in the rest of the tube, so that they still make exactly one circuit. This is the principle still used to make particles loop around the Large Hadron Collider. There too, they enter at low energies and are accelerated a tiny bit with each cycle. At the same time, the magnetic field is gradually increased to the maximum strength. At that point, we have to stop increasing the energy of the particles, because otherwise they would fly off.

When using cosmic rays, you had to wait and see how long it takes for particles to hit your apparatus and how much energy they have, but particle accelerators give you precise control of those factors. Since Oliphant's invention required nothing but a magnetic field in a small vacuum tube, it was much more stable than earlier accelerators, the first step on the path to a great leap toward far higher energies. Like those in the Large Hadron Collider in the early twenty-first century.

Learning how to cause collisions between two beams of accelerated particles (instead of aiming a single beam of high-speed particles at a stationary plate, as we had before) was the last big step in increasing the energy of the particle collisions. That's how the Large Hadron Collider works: two beams of protons are accelerated, and when they reach maximum velocity, they're both fired at the same exact spot: the proverbial eye of the needle. Then there's almost no way the protons can avoid a collision, and it's around those collision points that we physicists construct our experiments. All the particles made in the collisions and their remnants fly outward, passing the different detector layers that we discussed earlier: the tracking detectors that record the path, charge, type, and velocity of electrically charged particles, and then the calorimeters that measure the particles' energy. By combining the information from the different layers and looking at the characteristics of the collisions, we can see whether new particles were created.

Over a twenty-year period, we discovered lots (and lots) of particles with our accelerators: The Λ particles, the Σs, the Δs, the ρs, the ηs, and so on. The properties of each particle were collected and recorded in detail in the "big book," the *Particle Data Book*. It was an interesting but also a confusing time. Some people have described the work involved as "stamp collecting," a joking reference to the somewhat arrogant remarks some physicists make about other fields when new observations aren't accompanied by any "real" new fundamental insight. The physics of this period, full of confusion and data gathering, may seem boring, but it was tremendously important. As usual, the muddle continued until the moment when someone noticed the underlying patterns that made all the measurements and observed patterns fall into place. We sometimes forget the very similar earlier period of classifying atoms, which eventually led to the insight that, despite all their diverse properties, they are all made from the same three building blocks: protons, neutrons, and electrons. Looking back today, we can see the logic of the system, but that insight came only after years of experimenting and probing for patterns.

Once physicists had discovered a large number of particles, it was clear that the new particles could be classified into two groups by weight. The heavy particles, which weighed about as much as a proton or neutron, were named *baryons*. But there was also a group of particles, like pions, with greater mass than an electron and less than a nucleon. This latter group was dubbed the *mesons*. Within both groups, we saw surprising things. For example, some particles had *exactly* the same mass but different electrical charges. Take the four Δ (delta) particles (Δ, Δ^0, Δ^+, and Δ^{++}), which each weigh 1,232 million electron volts, or MeV (a little more than a proton, which weighs 938 MeV). Although they are all the same to the nuclear force, the electromagnetic force treats them completely differently, because each one has a different electrical charge. There were two other sets of that kind: the three Σ (sigma) particles, which have a mass of 1,385 MeV, and the two Ξ (Xi) particles, with a mass of 1,533 MeV. In each of these groups, the particles shared not only the same mass, but also another property known as "strangeness." We don't need to get into that property here, except to mention its influence on how, and how quickly, the particle decays.

Now have a look at the pattern on the left in the next figure, keeping in mind what you just learned about the mass of each group. You can see that the particles in each row are about 150 MeV heavier than the ones in the row above and lighter than the ones in the row below. And the heavier the particle is, the fewer particles there are in the row. Asked whether this pattern is complete, you would probably guess that one particle is missing at the third corner of the triangle, with a mass of 1,680 MeV (another 150 MeV heavier) and an electrical charge of -1. No one had ever seen such a particle, but it seemed logical to suppose that it existed, since it would complete the pattern that was clearly emerging.

In the 1960s, Murray Gell-Mann and George Zweig each independently solved the puzzle. All the patterns (the masses and the properties) could be explained if you assumed that all the baryons and mesons observed in our experiments were made of three smaller building blocks: quarks, which were named the up quark, the down quark, and the strange quark. These quarks had electrical charges that were, respectively, $+\frac{2}{3}$, $-\frac{1}{3}$, and $-\frac{1}{3}$ of the value of the electrons. If you assume that the up and down quarks are equal in mass, while the strange quark weighs 150 MeV more than they do, then right away you can explain the difference in mass as you go from one row to the next. In each row, one strange quark is added to the combination of quarks that forms the particle. The best thing about this model was that all the observed particles fit into it. If you have three quarks to combine, then exactly ten combinations are possible, and those combinations correspond precisely to the particles in the pattern. Or at least, we had observed nine of them. The tenth, the Ω^- (omega minus) particle in the bottom corner, had not yet been seen, but it was predicted by the quark model.

When that particle was discovered that same year, with all the properties as predicted, it was a triumph for the quark model. From that moment on, the pion particle and the nucleons (protons and electrons) lost their status as elementary particles. Instead, they were "merely" two of the many composite particles made of quarks. The proton was still special, because it was stable, while all the other particles made of quarks were very short-lived.

In hindsight, it's easy to grasp the solution, but hindsight is always 20/20. The question is always: who will be the first to put together the pieces of the puzzle to see the bigger picture? You might think that amazing discoveries like these would have everyone cheering from the sidelines. They represented

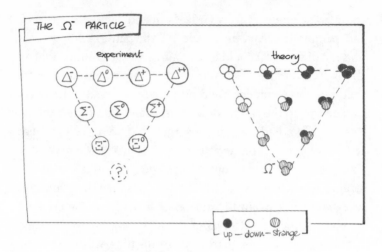

a breakthrough on the same scale as that moment in the early twentieth century when the number of basic building blocks was reduced from ninety (the number of elements) to three by the discovery of the nucleons and electrons. Likewise, the dozens of heavy particles had been reduced to a handful of quarks.

Even so, there were critical voices. Although the quark model explained the measurements perfectly, the question remained: were quarks merely a mathematical trick, or did protons really have a substructure? No one had ever seen a single quark in an experiment. And there was another problem: the particles at the three corners of the figure are each composed of three identical quarks—three quarks of exactly the same type, with exactly the same properties, like mass, spin, and spin direction. Our theory told us that was impossible. Why? Well, Pauli's exclusion principle, which we ran into in an earlier chapter, states that different particles in a system can never have exactly the same properties. That principle explained the structure of electrons in atoms and had become a fundamental requirement in the quantum mechanical description of elementary particles. Three identical particles in a combined state? Unthinkable!

The first problem—namely, showing that a proton really is made of quarks—was solved with the same trick used by Rutherford in 1910: making particles that were much smaller than a proton and bouncing them off the proton to see if we could describe its shape. Was it solid through and through, like a billiard ball? Or did it have an internal structure, like a walnut? More specifically, did it consist of three smaller balls at its core? In 1968 we finally managed to accelerate particles to a high enough energy to make them smaller than the proton itself, so that we could use them as microscopes to look inside the proton. It was immediately clear that the proton does have a hidden substructure.

The solution to the second problem, the more theoretical puzzle of the three identical particles in a composite system, was also simpler than expected. If we wanted to save Pauli's exclusion principle, we had to find some difference between the three identical-looking quarks. The similar problem of identical-looking electrons had been solved when young scientists in Leiden thought up the previously hidden property of spin. We could solve the quark problem the same way, by concluding that they had some extra property we had missed before, with a different value for each of the three quarks in the larger particles. This new property was named "color" and assigned three values: red, green, and blue.

The term "color" may be confusing, because we associate it with a property in our everyday world, but it works the same way as the property of electrical charge, which also has a limited set of values $(0, \pm 1e, \pm 2e, \ldots)$ or spin, which has two possible values ($\pm 1/2\hbar$). In short, we discovered a property of quarks we hadn't known about, but that made them satisfy Pauli's exclusion principle. Each quark came in three versions, and the three quarks in a composite particle were therefore not identical but each had a unique color. So quarks existed and were the new front line in the search for the elementary building blocks of nature.

Three other quarks were eventually found alongside the up, down, and strange quarks. They were much heavier than the three light ones, so it took a while longer before they showed up in collisions in particle accelerators. The charm quark was discovered in 1974, the bottom quark in 1977, and, finally, in 1995, we observed the top quark, which is 180 times as heavy as the proton and still the heaviest known elementary particle. The six quarks were divided into three pairs, like the electron-neutrino pair and its two counterparts. The up and down quarks formed the quarks in the first family. The charm and strange quarks made up the pair in the second family, and the bottom and top quarks were in the heaviest family, the third. In other words, the quarks followed the same pattern as the electron-like particles and neutrinos. Why? It's a complete mystery.

Antimatter: Its Prediction, Discovery, and Applications

The twelve elementary particles we've discussed form a beautiful pattern and are all we need to understand the world around us. But we weren't done yet. In fact, those twelve particles are only half the picture. We're missing one last magical ingredient: antimatter. The term is often used (especially in science fiction novels) for an exotic, mysterious, and deadly substance, and few ideas are more fascinating to the general public. Meanwhile, to particle physicists, antimatter is completely ordinary.

We've found that each elementary particle has a twin that's identical in almost every respect—the same mass, the same lifespan, the same mode of decay—but with exactly the opposite electrical charge. As particle physicists see it, antimatter therefore makes up 50 percent of all elementary particles.

Yet because antimatter doesn't occur naturally on earth, or in the rest of the universe, the general public sees it as something magical. The best-known example may be the bomb used to destroy the Vatican in Dan Brown's book *Angels & Demons*, made from a quarter gram of antiprotons. Although that's possible in theory, the reality is very different. Antiprotons can only be made and isolated in large laboratories like CERN, under controlled conditions. To make a quarter gram, it would take a few hundred million years. Not very realistic, in other words, but writers don't let that stop them. Experimenters at CERN have even managed to make antiatoms, combining antiprotons and antielectrons to form antihydrogen. A fascinating line of research, and an amazing technological feat. At first sight, the study of antimatter might seem like the ultimate academic pursuit: "OK, so it exists, but what's it good for?" Yet even though antimatter occupies one of the weirdest corners of the scientific world, it is used every day in major hospitals to examine tumors. You may have heard of a PET scanner. There's probably one in a hospital near you. But what is antimatter, exactly, and what is it used for?

When quantum mechanics was still in its infancy, the Englishman Paul Dirac was the first to succeed in combining Einstein's theory of relativity with quantum mechanics. He found a formula with which he could predict the movements of the electron in that strange quantum world (an equation of motion). When you study a natural phenomenon, whether it's a falling apple or communication between bees or people, it's crucial to combine everything you know into what's called a model, which gives you a clear picture of all the factors at play. When you've figured out how the different parts of your model interact, not only will you know just what to expect every time you run into the same situation, but you'll also be able to make predictions about completely new situations. In physics, a model is often a mathematical description of the different forces at work, and it enables you to answer any possible question about the system. In Chapter 4, I will discuss in detail the main formula in the Standard Model. It covers not only the elementary particles, but also the forces that make them attract and repel each other and take on new forms.

If you picture an electron as a kind of tiny bullet moving through space, then you might think you can apply the same centuries-old formulas used for a swinging pendulum or a thrown ball. But the discovery of quantum mechanics and electron spin made it clear that elementary particles obey different laws. Dirac's equation of motion, known simply as the Dirac equation, proved to be a very elegant, concise formula that, despite its simplicity, correctly described the complex behavior of electrons.

When Dirac looked more deeply into the consequences of his equations, he found something strange hidden in the formula. The model predicted that alongside ordinary electrons, there should also be electrons with negative energy. Of course, that was ridiculous. How can anything have negative energy? It seemed like a classic example of a "wrong solution."

Strange situations like this come up every once in a while in physics. If you drop a marble from the top of the Dom cathedral tower in Utrecht, then the formulas you learned in high school tell you it should take 4.5 seconds to reach the ground—or, strangely enough, *minus* 4.5 seconds! You normally sweep this "weird" solution under the rug as quickly as possible, but in Dirac's case, that wasn't so easy. He found a way to get rid of the negative-energy particle, but it would mean that an electron was moving backward in time. From the frying pan into the fire, in other words.

But theoretical physicists have all sorts of tricks up their sleeves, so he found yet another interpretation. Instead of an electron moving backward in time, you could picture it as an electron with a positive charge: an antielectron. If you think this is strange . . . you're right. Every physics student who sees this step in a quantum mechanics seminar thinks it's a bunch of hocus pocus, but soon learns to work with it. Dirac's claim, in the 1920s, that there must be something like an antielectron—we call it a *positron*—was one of the bravest predictions in the history of theoretical physics. Why brave? Well, for instance, the positron would have the same weight as the electron, and particle physicists had never seen an antielectron in all their hundreds of experiments.

Theoretical physicists still make predictions about the existence of strange particles every day, but 9,999 (or maybe 10,000) out of 10,000 predictions turn out to be complete hogwash. Dirac's antimatter was one of the first predictions of this kind, but this strange prediction turned out to be right. Pretty soon after he proposed the existence of the positron, the first collision experiments with cosmic rays actually revealed "positive electrons." So Dirac's antiparticles were real, and today we even know that every particle has a corresponding antiparticle. That means there are not just twelve elementary particles, but twenty-four. Thanks to the laws of nature, that also applies to composite particles. So if you can make a stable particle (a proton) out of two up quarks and a down quark, then there must also be a combination of two antiup quarks and an antidown quark (the antiproton). It couldn't be created until 1955, when particle accelerators were finally powerful enough to squeeze together enough energy, but once we had the technology, we soon discovered the antiproton.

After all the unsolved problems we've seen on our journey into ever smaller structures, you won't be surprised to hear that antimatter, too, still has its mysteries. Everything on earth is made of matter and not antimatter. In fact, throughout the universe, antimatter seems to be missing. But if the laws of nature are so symmetrical, why is there matter in the universe at all, and why didn't all the matter and antimatter particles destroy each other when the universe began? One of the exceptional aspects of particles and antiparticles is that they can meet and annihilate each other. This is the problem when we produce a positron in one of our experiments. Unless it's created in a perfect vacuum, the positron will soon run into an electron. After all, electrons are in every atom on earth, so they're in the gas in the detector and the materials of each detection layer. As soon as the positron and electron collide, they destroy each other and produce two flashes of light. So what happened to all the antimatter? Was there a little more matter than antimatter when the universe was born? Or were there other mechanisms that made all the antimatter disappear, leaving only matter? The Standard Model gives us ways to account for a slight asymmetry, but those are not nearly enough to explain the vast difference we see in our universe. So what's the solution? Nobody knows. That's exactly what makes this research so exciting. Eventually, it may tell us why the universe contains *anything at all.*

Even though antimatter occupies the most surreal part of the scientific spectrum, we've found a surprising way to use it in human society. Not for building bombs like Dan Brown's, but for hunting down tumors with a positron emission topography (PET) scan. The two flashes of light that a positron produces when it meets an electron can be used in a PET scan to locate tumors.

The first step is to inject the patient with a radioactive substance: radioactive atoms attached to large molecules (often sugar). These molecules are carried by the bloodstream to the tumor and accumulate there. How, exactly? That's medical science, and I won't get into the details. The radioactive atom is what we call a *beta-plus emitter*: when it decays, it emits not an electron (beta radiation), but a positron. In other words, antimatter. After that positron is released in the radioactive decay process, it soon runs into an electron in the body. As the two particles annihilate each other, they create two photons, which shoot straight through the body in opposite directions. Outside the body, they are recorded by ordinary cameras. So if you see two photons at approximately the same time, you know that there was a positron-electron collision along the line between the two cameras. That must be the location of the radioactive atom, and therefore of the tumor. Through very accurate measurement of the arrival time of each separate flash of light, you can even figure out *where* on that line the annihilation took place. Because the patient is injected with many, many atoms of radioactive isotopes, and the two particles always fly off in random (but always opposite) directions, you can put together a three-dimensional image of the tumor. Using antimatter in hospitals to track down tumors . . . who would ever have imagined it?

The Elementary Particles: A Complete Set

After our initial confusion, we learned from the patterns we discovered that many of the new particles were made up of two or three quarks. The number of *elementary* particles

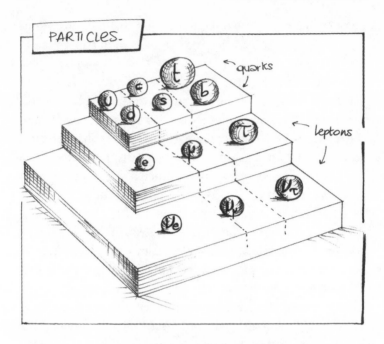

turned out to be rather small. Leaving aside antimatter for the moment, there are twelve: six quarks and six leptons.

The six quarks are the up quark, the down quark, the charm quark, the strange quark, the bottom quark, and the heaviest elementary particle, the top quark. Single quarks are not found separately in nature, but combinations of two or three can form particles (mesons and baryons, respectively) known collectively as hadrons. The most familiar hadrons are the proton and the neutron, which are both combinations of up and down quarks. Together they form the atomic nuclei of all substances on earth and throughout the cosmos.

There are six leptons: the three with an electrical charge (the electron, muon, and tau particles) and their partners, the three neutrinos (the electron neutrino, the muon neutrino, and the tau neutrino).

In the fifty years from 1920 to 1970, great progress in the design and power of particle accelerators and detection techniques allowed us to penetrate ever further into the jungle of elementary particles. Besides the nucleons and electrons that make up atoms, we also found a ghost particle, and the nucleons turned out to be made of quarks. This family of four and its two extra copies gave us twelve elementary particles, which could combine into many heavy, short-lived composite particles. And just to make things even stranger, each particle turned out to have an antiparticle. As you can see, even as we gained deeper insight into the building blocks of nature, they remained as mysterious as ever. Why are there three families of particles (and not just one)? What happened to all the antimatter? And why do particles have such different masses?

We discovered not only these particles, but also the laws that determine their behavior depending on their properties. Why are some particles produced more often than others? Why can an electron and positron meet and annihilate each other, while a positron and muon cannot? We eventually learned this set of laws, nature's rulebook, and we now call the whole collection of particles and rules the Standard Model. The formulation of the rules is deceptively simple to write down, considering the richness and depth of the phenomena it describes. Now that we're acquainted with most aspects of that theory, we can see that everything seems to stem from a small set of mathematical symmetries. All the complex behavior we observe turns out to be based on a small number of sturdy pillars whose origins are a complete mystery to us. What's so deep about that observation, and how does the Standard Model work, exactly? We'll soon see that the only way to keep the whole edifice of particle physics standing is by adding one final pillar: the Higgs boson.

4
Forces in the Standard Model

Now that we've identified the elementary particles that form the building blocks of all matter in the universe, the next step is to understand how those particles communicate with each other. We previously saw that particles can attract and repel each other, depending on their mass and electrical charge. We also learned that quarks are trapped inside stable particles like protons. But what's going on when a particle meets its antiparticle? How are new particles produced in collisions? And how do nuclei decay? What laws govern nature in these processes, and what makes the "forces of nature" work?

It took years to construct a comprehensive theory in which every particle and every force has its place. We pulled it off by studying each separate phenomenon in detail and then putting in a lot of hard thinking. When the dust settled, we realized that particles actually have only four unique properties that determine how they behave. Those four properties, combined with the rules of play, explain how particles interact and, in the process, shape our world. The four forces of nature (and the corresponding properties, or "charges") are:

1. *Gravity (mass):* Describes how particles attract each other and cluster into planets and stars, and why an apple falls to the ground.
2. *Electromagnetic force (electrical charge):* Tells us how electrically charged particles attract or repel each other. It also explains most properties of materials.
3. *Weak nuclear force (weak isospin):* Makes particles change into other particles and is also responsible for radioactivity. Works only over tiny distances.
4. *Strong nuclear force (color):* Describes how quarks clump together into particles like protons and neutrons.

These forces form the basis of the Standard Model and provide a solid foundation for our experiments and our quest for new phenomena.

After years of effort, trial, and error, the pieces of the puzzle finally fell into place in the 1960s and 1970s. Gravity is still the odd force out, even today, but the other three (the quantum forces) turned out to have the same basic mathematical structure. Despite the huge differences between them, they could be understood as different examples of the same underlying process: particles communicate by exchanging what are known as messenger particles or force carriers. Along with the building blocks we covered in the last chapter, these force carriers and the laws governing them were combined into the famous Standard Model.

Although the mathematical structure of this model was extraordinarily beautiful, it had a serious problem: it couldn't describe a world in which particles have mass. That's more than unfortunate—it's the kiss of death for a theory that aims to describe nature. After all, particles *do* have mass. To save our beautiful mathematical structure and yet assign mass to particles, we had to introduce a slight imperfection into our otherwise perfectly symmetrical framework. That final adjustment, proposed by Peter Higgs and his colleagues, completed the foundations under the Standard Model. But it came at a high price. If their ideas were right, then there had to be one more particle hidden in our world: the Higgs boson.

When exploring the world of elementary particles, we particle physicists use virtually the same strategy as children trying to understand the world. The knowledge that children pick up from their parents includes not only practical skills but also the kinds of interpersonal rules that apply in any society. These rules aren't simple. They depend on hundreds of factors, and to make things extra complicated, they also depend on the social context. In nature, an apple always falls down, but a person's response to an action depends on the social setting. A book club in a wealthy metropolitan neighborhood has a very different culture from a bowling club in a suburban town, and a soccer supporters' club has different rules of interaction than a golf clubhouse. As soon as you've learned (sometimes through bitter experience) which factors and rules are relevant to your own life, the game unfolds according to regular patterns. No more unpleasant surprises. You can rely on your general model of society even in situations that you've never experienced before.

To discover and understand the laws of nature, scientists take a similar approach. They too are in search of the basic rules of the game. The big difference from all those tricky social interactions between people is that the laws of nature are unchanging and implacable: in any given situation, nature will always react exactly the same way. And what makes scientists different from most people, who enjoy the wonders of nature without a second thought, is that scientists try to strip down nature to the bare essentials. They create situations in which they isolate just a few properties in order to get to the heart of the issue. Whether it's a falling apple or communication between bees or human beings, their question is always, What are the underlying regularities? I have to admit it's a never-ending quest, going from one *why* to the next. Yes, of course an apple falls when you let go of it. Anyone can see that. But *why* does it fall, how fast does it fall, and does the shape of the apple matter, or its weight? As soon as you know *how* nature works, the next step is to figure out *why* it works that way.

Physicists describe how nature works in formulas: the most compact, bare-bones way of writing down the rules, without a lot of fluff. Formulas describe how nature works, as far as we know today. If a new phenomenon is observed that contradicts the rules we've figured out so far, then we'll have to change them. It's that simple. In that continuous process of improvement and adjustment, sometimes a scientist arrives at the same rules from a different angle or a new basic principle. Major paradigm shifts like that happen only once in a great while, but when they do, they almost always provide a wealth of new insights.

Most people are not used to "reading" formulas. So I generally avoid using them, and you don't need them, strictly speaking, to grasp the essentials. That said, I'll do my bit to fight one of the most stubborn fears in the world today: formulaphobia. After all, it was thanks to these formulas that we learned that empty space wasn't empty at all and predicted it should be filled with what we call the Higgs field. To show how we arrived at that prediction—namely by making a paradigm shift, via quantum versions of familiar forces and new properties, to the mysterious properties of the nuclear forces— I'll lay out the steps that led to deeper insights as we discuss each individual force.

Gravity, the Misfit Force

Before I zoom in on each of the three forces in the Standard Model, let me start with a few facts about gravity. Even though it's the best-known force of nature, you won't find it in most books about particle physics. At most, a brief footnote might tell you that gravity is so much weaker than the other three forces that it plays no serious role in the world of the very smallest particles, so we can stop talking about it. In any case, gravity doesn't fit into the Standard Model, because we still haven't found a theory that unites gravity with quantum mechanics. So why should we waste time discussing it? It's true that gravity matters only in the world of the largest objects, where matter clumps into planets and stars.

But I'm starting with gravity anyway, to show you a few things about how we scientists use formulas to penetrate to the sometimes hidden essence of interactions, and how new ideas emerge that can change our perspective completely.

The famous law of gravity proposed by the English physicist Isaac Newton in the late seventeenth century states that two particles with mass attract each other. That may sound simple, but consider this: even today, we still can't explain it. *Why* should particles move toward each other? Even so, once you've accepted it as fact, then it's logical that an apple always falls downward, and that whenever we humans jump, we always fall back to the ground. In both cases, there's something heavy pulling on us with an inescapable grip: the earth itself. If you study the behavior of this force in greater detail by experimenting with it, then after fiddling around for a while, you'll reach the same conclusion as Isaac Newton, whose name remains linked to this force. He observed that the force (F) between two particles depends on the mass of those particles (m_1 and m_2) and the distance (r) between them squared. The strength of the force is also determined by an unchanging factor, the gravitational constant (G). The formula looks like this:

$$\text{Gravity until 1915: } F = G\frac{m_1 m_2}{r^2}$$

The heavier the particles, the harder they pull at each other, and if the particles are twice as far from each other, their pull on each other is four times weaker. That's pretty much all there is to it. The great thing about this law is that *all* the regularities of gravity are tucked inside it, and you can use it to predict things. For example, the equation can help you to understand how and why the moon revolves around the earth, and where a tennis ball will land when you return it. Another interesting thing appears when you combine this with a more general law of motion that Newton also devised, the one that says that a force is mass times acceleration. The combination of these two laws tells you that when you compute the acceleration of a particle—in other words, when you think about how a particle will gain or lose velocity under the force of gravity—the mass cancels out. What does that mean? Well, this corrects one of the biggest misunderstandings about nature, namely that more massive objects fall faster than light ones.

More concrete: Newton's formulas tell us that a tank falls at the same speed as a frog. A lot of people find that hard to believe, but it's true.

The laws of nature are especially useful because they're universal, applying not only on earth but throughout the cosmos. So this formula can just as easily predict how an apple will fall if you drop it on the moon. If the moon is about eighty times as light as the earth and its radius is four times as small, then its pull on the apple will be about five or six times as weak as the earth's. That's how astronauts can make such giant leaps, and why Neil Armstrong and Buzz Aldrin didn't need a big rocket to get off the moon when it was time for them to leave. It's pretty helpful to know that kind of thing in advance.

Newton's formula for gravity works perfectly, and for hundreds of years there was not the slightest hint that it might be incomplete. Sure, we couldn't—and still can't—answer the simple question of *why* particles attract each other, but what made us rethink Newton's formula, in the early twentieth century, was a new insight from the mind of Albert Einstein. He caused a paradigm shift in the way we look at gravity. In his famous theory of relativity, Einstein concluded that space and time were not as static as people had thought; instead, they were thoroughly tangled up with each other. He also made the even more startling prediction that massive objects distort the space they're in. Expressed as a formula it looks like this:

$$\textit{Gravity after 1915: } R_{\nu} - \frac{1}{2} R g_{\nu} + \Lambda g_{\nu} = \frac{8\pi G}{c^4} T_{\nu}$$

This is where I lose you. I know what you're thinking: "Newton's formula was one thing, but now he's gone too far." Fair enough. I just showed you a bunch of Greek letters and symbols without telling you anything about what they mean. Since you can't read the formula, you can't see its point or its value. But this one line, the distillation of Albert Einstein's thoughts on gravity, is a gold mine of information. That information transformed our worldview *completely*, and it took several years to extract all its gems of insight. So now it's time to venture deeper into the language of mathematics and formulas. What makes them so confusing? How can you keep your fear of formulas under control? And how can you gain a feeling for the beauty that lies within them?

Scatter a handful of Greek letters on paper, add a few strange symbols, and right away lots of people start to groan and sweat. That's too bad, because mathematics is meant to be a compact way of communicating the framework of your thinking, stripped of all the extras. I've heard the same old advice so many times, whenever I have to write an article or give a talk for a nonscientific audience: "Use as few formulas as possible and don't get too abstract." I get the idea. The language of the natural sciences is the language of mathematics, but because few people have mastered that language, it often looks complex and intimidating. Still, there's a fairly simple analogy that tends to put people at ease, in my experience. So here goes.

If I give you a piece of paper with a string of Japanese characters and ask you whether it's a clause from the Japanese tax code, a line from a poem by Pablo Neruda, or a sentence from the children's book *Miffy at the Seaside*, you'll be stumped. How can you answer that question? Japanese is known for its complicated writing system, and you've never learned the language. But at the same time, you know that millions of Japanese schoolchildren would have no trouble answering my question and translating the characters for you. In the case of *Miffy*, a simple translation will suffice, but if it's a line from a Neruda poem, that won't be enough. A Japanese child could translate the words for you but won't recognize the message hidden between the lines of the literal translation, the references to literature and art, or the emotions described, and therefore won't be able to dig through the layers of meaning to the underlying message hidden in the poem.

Formulas are like those lines of Japanese. They come in all shapes and sizes, ranging from the shallowest trivia to statements of unfathomable depth and unexpected insight—such as Einstein's formula for gravity. And oddly enough, there's more to Albert Einstein's mathematical poem than he himself ever imagined.

To understand Einstein's formula in all its facets, you have to study physics for five years (at least) to get a handle on the math. And even then . . . Like that Japanese child, I could translate it literally for you and add a few words of explanation, but to reach the treasures that lie buried in its utmost depths, we need some help from the grandmasters of formula reading: the formula whisperers. In other words, theoretical physicists.

Imagine that nature is a play unfolding on a vast stage in front of us. Einstein's great insight is that the stage, space-time itself, is not stationary, but dynamic and plastic, and that it can be deformed. He taught us that we should see the stage not as a collection of wooden boards, like the ones in a traditional theater, but rather as a stretched elastic surface, like a trampoline, that bends underneath an actor or another object. The most interesting thing is that the formula describes both the actors on the stage (the right side of the formula, the particles and energy) and the stage itself (the left side of the formula, what he called space-time). The equals sign tells us that the stage and the actors are related to each other, and his formula allows you to predict exactly *how* the stage is curved and deformed as an actor crosses it. It makes for an interesting play: whenever two actors come near each other, the deformed stage pushes them toward each other. That's the insight that Einstein reached, and passed onto us, about *why* two particles attract each other. We no longer have to say that two particles simply attract each other according to Newton's laws—period! Particles attract each other because they slide toward each other on the curved stage. This analogy is even powerful enough to explain some other properties of gravity. For example, when two actors stand close enough together, they make a deeper pit than either of the two would separately and are even more likely to capture other actors. In the language of physics, heavy objects pull harder than light ones.

But many other wonderful things also lie hidden in Einstein's ideas (as expressed in his formula):

- **Black holes** A large enough cluster of actors will create a pit in space-time so deep that none of them can ever escape it (or each other). A pit so deep that even at nature's top speed, you can't get out of it. That's what we call a black hole.
- **Gravitational waves** If two actors spin in circles around each other, then the deformations cause ripples that spread through the entire trampoline. As small as those ripples are, they can be felt anywhere on the stage. Tiny gravitational waves of this kind, caused by two black holes or neutron stars revolving around each other very far away from the earth, were observed for the first time in late 2015. In the words of one of the discoverers, Jo van den Brand of Nikhef, "We physicists have gained a new sense and fantastic new insights."

- **Space-time** In spots where the trampoline is stretched, time changes too: it moves more slowly. It takes a lot of work to draw that conclusion from the formula. But we've actually observed that you can stretch time. A clock on a satellite runs a little slower than one on earth. If you don't keep that in mind, you can't use GPS to figure out where you are on earth.

While cosmology is the branch of science that looks at the stage itself, space-time, we particle physicists are mainly interested in the actors, the elementary particles. What I find most fascinating is that the search for the laws describing the world of the very smallest particles, the three quantum forces, has ultimately taught us something about the whole stage. We've found out that space-time is filled with an energy field, the Higgs field. Without that field, the actors would have no mass at all, and the universe would not be full of stars and planets. Then that beautiful play we all enjoy when we look up on a clear night would have been a lot duller.

In summary, we now believe that two particles with mass attract each other because space is curved. And it's the particles themselves that curve it. Gravity is much weaker than the other forces of nature, and really only matters on a large scale. In the world of elementary particles, it plays no meaningful role, and it won't come up again in this chapter.

Gravity

Property: mass
Particle: none (or the graviton, according to some new ideas)

Behavior: always attracts, very weak
Bizarre feature: black holes

But if gravity doesn't play a meaningful role in the world of elementary particles, what forces *do* we find at the foundations of nature? And how exactly do they work?

The Three Fundamental Forces of the Quantum World

The search for rules in the world of tiny particles raised a lot of fundamental questions. Could we figure out how two electrically charged particles that come close to each other know of each other's existence? Could we learn exactly what happens as they approach each other, and what makes them turn away from each other? And could we calculate how that works? Would we be able to determine how new particles can be created in particle collisions, and could we find the reason that some forces operate over large distances and others only at close range? The answer to all these questions is "yes." In fact, we now know that despite the huge differences, all forces have the same structure. And we'll be better prepared for a detailed look at the peculiarities of each individual force if I first unveil the big secret: the basic concept behind all three forces.

We've found that particles communicate by exchanging what we call force carriers. These force carriers communicate the particle's properties to the outside world. For instance, electromagnetism communicates the property of electrical charge. The weak and strong nuclear forces communicate two properties that I'm guessing are less familiar to you: *weak isospin* and *color*. Understanding these concepts is worth the effort, because doing so gives us the power to understand the results of hundreds of experiments in one fell swoop.

ELECTROMAGNETISM AND ADAPTING TO THE LAWS OF QUANTUM MECHANICS

At the dawn of the twentieth century, we knew of only one force other than gravity: electromagnetism. That was another success story: again, we had broken through to the essence of the force. In electromagnetic interactions, the behavior of the particles depends not on mass but on a different property: electrical charge. As in the case of gravity, the force between the particles rapidly decreases as they grow farther apart, and the formula looks very similar:

$$Electromagnetism:\ F = \alpha \frac{q_1 q_2}{r^2}$$

The force (F) is determined by the electrical charge of the particles (q_1 and q_2), the distance between them (r) squared, and a constant (α). The main difference between gravity and electromagnetism, aside from the strength of the force, is that while gravity is always attractive, pulling things together, electromagnetism can be either attractive or repulsive, pushing things apart. Different electrical charges attract each other, and like charges repel each other. The concept is so well known that it's even become a cliché in our everyday world: "Opposites attract." Just like gravity, electromagnetic force is four times as weak when the particles are twice as far apart.

I've been using the present-day term electromagnetism, but until the mid-nineteenth century, magnetism and electricity were two separate phenomena. It wasn't until the English physicist James Maxwell had captured every aspect of the two forces in his famous Maxwell equations that he was able to see them for what they were: two sides of the same coin. Until the early twentieth century, his famous equations and Albert Einstein's laws of gravity formed the basis of all known laws of nature. Of course, that was before the quantum revolution of the 1920s turned everything upside down.

When scientists probed deeper into the world of the very small, they found that the classical laws of nature didn't necessarily apply there. On a large scale, you can see an electron as an electrically charged marble, but on a small scale, you have to consider quantum effects. Then it makes more sense to describe the electron as a wave of a certain amplitude, with properties that are not all well-defined. It took a long time before anyone found a description of electromagnetism that was completely consistent with both the theory of relativity and quantum mechanics. But by the late 1950s, there it was: quantum electrodynamics. One of the leading thinkers behind this new theory was Richard Feynman, one of science's most captivating characters. For this masterstroke, Feynman—along with Julian Schwinger and Shin'ichirō Tomonaga—was awarded the Nobel Prize in Physics in 1965. You might describe the exact formulas and calculations involved in quantum electrodynamics as difficult—if you're fond of understatement. Or as Feynman once said to a journalist, "Hell, if I could explain it to the average person, it wouldn't have been worth the Nobel Prize."

One interesting part of this story about quantum electrodynamics is that Feynman found a graphic way of representing his equations: the pictures called Feynman diagrams. Non-physicists see them as schematic diagrams of interactions between two particles, but to particle physicists, they are recipes for very complex calculations. Just one of those calculations could cost you a couple of rainy Sunday afternoons. Fortunately, the essence of a diagram is easy to grasp, so you can think about the concepts without doing all that math.

In a Feynman diagram, time often runs from left to right. Particles are drawn as straight lines, and force carriers as waves. One side effect of the theory of quantum electrodynamics and its diagrams was to make it clear that a particle and its antiparticle can produce a force carrier if they're fired at each other. If that force carrier (the photon, γ) then falls apart into a different particle-antiparticle pair, then you've created new matter. You often hear in science fiction movies that matter and antimatter annihilate each other, creating a ball of energy from which new particles can emerge. We all know not everything in those movies is true, but in this case they're right.

To see the power of Feynman diagrams, take a look at this example, which describes exactly that type of process. A muon and an antimuon are produced after an electron and its antiparticle, a positron, meet and annihilate each other. In particle physics we describe this process as $e^+e^- \rightarrow \mu^+\mu^-$.

When an electron and positron meet and annihilate, they can temporarily create a (virtual) carrier of electromagnetic force, the photon, represented by the wavy line. The force carrier then "falls apart," turning into other particles. In this particular example, you see a muon and an antimuon, but they could just as well have been a quark and an antiquark. It's only in this type of annihilation process that photons fall apart like that, and the theory predicts that they will always produce both a particle and its antiparticle. That's not really such a strange prediction to make, considering that our theory is based on thousands of experiments, which collectively

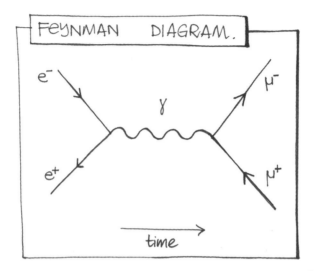

seem to show that this is how nature organizes the creation of new matter.

In short, we've succeeded in translating the familiar electromagnetic force into the world of quantum mechanics. Electrically charged particles can sense and communicate with each other by exchanging messenger particles, photons. Through this exchange of information, they learn of each other's existence and find out not only *whether* they should attract or repel each other, but also exactly *how*. One additional fact that's especially important in the world of elementary particles is that a particle and its antiparticle can meet and annihilate.

Electromagnetic force

Property: electrical charge
Force carrier: photon (massless)

Behavior: weaker at greater distances, infinite in
   range
Bizarre feature: a particle and its antiparticle can
   annihilate each other

One final point: the carrier of electromagnetic force, the
photon, has no mass of its own. That's why this force is percep-
tible over large distances. You might think that's just a detail,
but we'll see how crucial it is when we look at the weak nuclear
force. We'll need an explanation of why that force, unlike electromagnetism, is effective only over short distances.

THE WEAK NUCLEAR FORCE

We particle physicists are lucky: compared to the elaborate
rules of human social interaction, the rules in the world of
elementary particles are much simpler. Fundamental particles have only three essential characteristics, and each one
is linked to a different force. We've just seen how that works
in the case of electrical charge and the related force of electromagnetism. The other forces are the two nuclear forces.
They're less familiar, but that doesn't make them any less important to our story.

   1. **A new property: weak isospin** Warning: we're
      now entering a very strange part of the physics
      landscape. If the term "weak isospin" seems con-
      fusing, just think of some other property that
      has exactly two values and is easier to picture:
      for example, a switch that can be on or off. I'll go
      on calling it weak isospin, simply because that's
      the name we physicists use for it.

Although the weak nuclear force uses much the same method of communication as electromagnetism, there are three big differences in how force carriers are exchanged. Those differences give the weak force its unique nature.

As we noted earlier, particles that look the same may have a hidden property that distinguishes between them. Think of two job candidates who seem identical, but have different political views. There's a difference between them, but not one we can observe during the interview. We've talked about the discovery that electrons had a hidden property, namely spin. It came in two forms: spin up and spin down. That difference, previously undetectable, was what made it possible for two electrons to occupy the first atomic orbital together. We've also discovered other hidden properties, including one that's fundamental to the weak nuclear force: weak isospin.

You may not have noticed, but in classifying elementary particles, we have consistently grouped them into pairs. In the first family, for instance, the up quark was paired with the down quark, and the electron was paired with the electron neutrino. There's a good reason for that. Although the two particles in a given pair seem very different, they share a property: namely, weak isospin. This property has the same absolute value for each of the two particles, but like ordinary spin, it can be either positive or negative. If one particle in the pair has a weak isospin of $+\frac{1}{2}$, then the other has $-\frac{1}{2}$. The intimate connection between the two members of each pair will become clearer as we investigate the unique properties of this force.

2. **The weak nuclear force works only over short distances: heavy messenger particles** As theoretical physicists were developing their model of the forces at work in the nucleus, they faced a difficult dilemma. One constraint on their theory was that the nuclear force must be more powerful than electromagnetism. After all, electromagnetism pushes the protons in the atomic core apart, because they are all positively charged and packed close together. So the nuclear force must pull even harder to keep them together. The second constraint was that the nuclear force must be much weaker than electromagnetism outside the nucleus. If it weren't, then all the protons in your body would clump together into one big atomic nucleus. But we know that they don't. At first sight, these two constraints seem incompatible, but the problem was finally solved with one astonishingly simple adjustment.

If the force carriers had mass, that would explain why it was harder for particles to exchange them over larger distances. The best way to picture this may be as a watchdog on a chain. The dog can have a huge impact on your life, but only if you move closer to it than the length of the chain. As long as you make sure to stay farther away, it can only bark at you—not really

affect you. If we assume that the three force carriers of the weak interaction have mass, that
would imply that the force has only a limited
range. The particles are "chained up," so to speak.
This issue—the mass of force carriers—is what
would get the Standard Model into trouble and
put Peter Higgs on the trail of the Higgs mechanism. How do the force carriers acquire mass?

3. **Three force carriers, and how particles change
   as they communicate** Physicists found that the
   property of weak isospin was more complicated
   than its electromagnetic equivalent, which involves a simple positive or negative charge. Instead, weak isospin involves three different force
   carriers: the W^+, W^-, and Z bosons. We'll come
   back to the Z boson later, because it played a piv-
   otal role in the discovery of the Higgs boson. But
   it's the W bosons that make the weak nuclear
   force unique. So far, we've looked only at cases
   of communication between particles in which
   the particles themselves did not change. All they
   did was exchange information. Apart from the
   special case in which a particle and its antiparticle collide, all we've seen is that particles change
   their direction under the influence of attractive
   and repulsive forces, and change velocities. But
   an electron always remains an electron, and an
   up quark always remains an up quark.

   When we look at the weak nuclear force,
   this is no longer true, and the whole picture becomes more complex. When two particles with

weak isospin exchange a W boson, the particles themselves change. That's because the force carrier takes part of the weak isospin along with it. We might compare this to a world in which everyone is either rich or poor, and people show each other how much money they have by tossing their wallets to each other. The money in the wallet stands for the amount of isospin. When a rich person throws his full wallet at a poor person, he shows that he *was* rich . . . at the moment he threw it. But after the exchange, the situation is reversed: the rich person has become poor, and the poor person, rich. The W boson plays the same role as the wallet, but instead of money, it transfers weak isospin. This changes the nature of the particles involved: an electron becomes a neutrino, or an up quark becomes a down quark. This is not to say there are lots of possibilities. In our imaginary world, all the people were either rich or poor. Likewise, there are only two possibilities within each pair of particles: electrons and neutrinos can change into one another, and so can up quarks and down quarks.

There are several reasons that the weak nuclear force hardly ever turns up outside of particle physics experiments. The W bosons are so heavy and short-lived that particles have to come very close together to feel their effects. Yet our discoveries about the weak nuclear force have helped us to understand a number of processes—*truly* understand them. The best example is neutron decay, a vital part of radioactivity and the stability of atomic nuclei. We've known for a long while that a

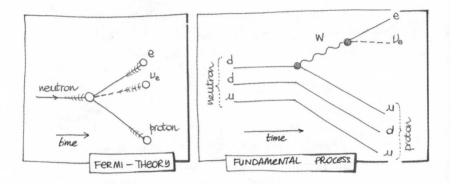

FERMI — THEORY

FUNDAMENTAL PROCESS

neutron decays after about nine hundred seconds, producing a proton, an electron, and an electron neutrino or antielectron neutrino. But the Standard Model and the weak nuclear force offer us a new way of looking at this process.

In this model, we see the neutron as consisting of one up quark and two down quarks. If one of the two down quarks changed into an up quark (which is possible, if it emits a W boson), then the result would be a particle composed of two up quarks and one down quark. That's the particle we identify as a proton. The W boson emitted in the process must decay into a separate pair of particles. Since it doesn't have much energy, the only possibility for it is to decay into an electron and an antineutrino. So we now not only have a more fundamental understanding of *why* a neutron decays, but can also calculate *how often* it happens. Yet another mystery solved.

To sum up, the weak nuclear force operates between particles with the property of weak isospin, and the range of the force is severely limited by the mass of the force carriers. No quantum force would be complete without at least one unique and mysterious aspect. In this case, it's the transformation of particles as they communicate.

Weak nuclear force

Property: weak isospin
Force carrier: W^+, W^-, and Z bosons
Behavior: stronger than electromagnetism, short
    range
Bizarre feature: can transform particles from one
    type to another

THE STRONG NUCLEAR FORCE

The final force is the strong nuclear force. This force operates only between quarks, because they are the only particles with the property to which it is sensitive: *color.* The term is confusing, because we use it in a different way in our everyday world. This quantum property known as color has only three distinct values: blue, green, and red. We could just as easily have called these values A, B, and C, or X, Y, and Z, but we happen to have settled on the three colors. Although the dynamics of this force are not easy to describe, they most resemble a rubber band stretched between two objects. When the objects are close to each other they can move freely, but the farther apart they are, the stronger the force that holds them together. The gluons, the carriers of the strong nuclear force, hold the three quarks together in a proton; they also hold together the lighter mesons containing two quarks. Since this property is called color, you can see why physicists call the theory of the strong nuclear force *quantum chromodynamics.*

There are a couple of unique twists: although the force increases as the quarks grow farther apart, it's also possible for a gluon to tear apart, like a rubber band that snaps when over-stretched. To be more exact, a gluon can split into two gluons or into a quark and an antiquark. The reverse is also possible: a quark and an antiquark that collide can form a gluon.

Strong nuclear force

Property: color (blue, green, red)—only for quarks
Force carrier: gluon—massless
Behavior: stronger as distance increases, like a rubber band
Bizarre feature: a gluon can split into two quarks or two gluons

If you delve into the predictions in detail and try to do the calculations for yourself, you'll soon discover that the mathematics of the strong nuclear force is very difficult to master. Although the mathematical structure of the force is similar to that of the two other forces at work in the nucleus, many of the calculations run into an impenetrable mathematical wall for which we have no solution. A few people have tried to break down this wall by developing new mathematical techniques or approaching the problem from a different angle. We've made a few dents, but we haven't found a way through it yet.

The Standard Model Formula and the Melody of Nature

Now that we've seen the full lineup of forces, the next step is to combine all the rules into one neatly organized model, so that we can predict exactly what will happen whenever any two elementary particles meet. This model is the large, comprehensive formula on the left in the next figure. It looks awfully impressive and daunting to the untrained eye—and to the trained eye, too! Earlier, we saw that formulas are like Japanese poetry: you have to learn how to read them in order to grasp the rhythm and discover any concealed layers of meaning. To make the idea even less abstract, I'd also like to compare them to sheet music.

Listening to music is something anyone can enjoy, but music is translated onto paper in the strange language of musical notation. Many musicians can hear the music in their heads when they read a score. It's funny that no one complains about musical notation being too abstract, even though to people like me, who can't read music, it might as well be Chinese. All I see is a thick forest of black dots and lines—complete gibberish. I couldn't tell the difference between, say, Bob Dylan and "Chopsticks" if my life depended on it. And I would never recognize a wrong note hidden in the sheet music, even though a professional musician could pick it out in no time. For a reader with an untrained eye, the melody remains hidden.

The same thing applies to mathematical formulas expressing problems in physics. Scientists are used to "looking through the formula" and seeing the beauty. They can tell right away if the wrong Greek letter was accidentally used, or if a plus sign was mixed up with a minus. Just as there's a difference between "Chopsticks" and a Rachmaninov sonata, there are many different levels of abstraction and complexity in physics. The formula for the motion of an elementary particle is a whole lot more complicated than calculating the movement of a swing in a playground. That stands to reason. But both calculations are based on a similar method: first you take a good look at which properties are important, and then you try to capture the phenomena you see in a mathematical expression. After that, you look for an underlying structure.

Standard Model formula   sheet music

Modifying the theory of electromagnetism so that it applied to the world of elementary particles was an unparalleled achievement, for which Feynman and his collaborators received a Nobel Prize. Besides a formula describing how individual particles move through space, they needed to come up with the idea of force carriers, which link the freely moving particles in very specific ways, enabling them to communicate with each other. To make things even trickier, the force carriers also had to meet strict conditions so that they would resemble the particles we had observed in our experiments. It was a complicated puzzle, but in the end everything fit together neatly—a triumph of theoretical physics.

When the formulas were examined in more detail, one scientist discovered that you would end up with exactly the same formulas if you started with a single charged particle whizzing around in empty space and imposed a few extra requirements—in this particular case, the condition that the particle would go on moving in exactly the same way even if you had the freedom to change the particle's wave function at every point in space and time according to a particular recipe. That's odd, because a physicist who looks at the natural laws describing the motion of an individual particle can easily see that if you made a change of that kind, you wouldn't end up with the same equation of motion at all. But here's the kicker: if you *really* want the motion to remain unchanged, in spite of that extra freedom, then there's a loophole in the theory that can do the job. But it comes at a cost, and has some serious strings attached.

To make it all work, you'd have to add another particle to your theory. So besides the particle moving through space, there would have to be another particle in nature. And not only that, but that extra particle would also need to have a number of very specific properties *and* be capable of communicating with the original particle. Miraculously, the extra particle you'd have to invent would turn out to have just the same properties as a known particle, namely the photon, the force carrier of quantum electrodynamics. As a matter of fact, it *was* the photon—in every detail. Wow, talk about strange: the force carrier, namely the photon, and the way it interacts with other particles didn't have to be "invented," but arose "automatically" from a seemingly bizarre mathematical condition on a single particle moving through empty space.

We call this extra requirement a symmetry condition, because things remain unchanged even if we make a particular modification. The new particle you have to introduce to keep everything consistent is a type you've already learned about—we called it a force carrier—but it's also known as a gauge boson. The term "gauge" refers to the type of symmetry that we're postulating: *local gauge invariance*. In writing these last three words I've managed to share with you one of the most complicated terms in particle physics! Of course, that doesn't mean we now understand the origins and existence of forces. After all, what is the logic behind the existence of this symmetry?

Things really got exciting when we found out that the weak nuclear force, with its three force carriers that interact in complex ways, could also be reduced to a similar type of symmetry requirement: a more complicated form of symmetry, which required not just one force carrier to keep everything symmetrical, but three: the W^+, W^-, and Z bosons. And what about the strong nuclear force? Yes, that one also turned out to "arise" from a symmetry condition. And it required eight force carriers: the gluons.

So the whole complex framework of physics at the quantum level could be reduced to three distinct symmetry requirements. In terms of sheet music, these symmetries are the underlying melody of nature. In the language of theoretical physicists:

The melody of nature: $U(1)_Y \otimes SU(2)_L \otimes SU(3)_C$

Thanks to this melody, we can reduce the very lengthy and complex formula describing all particles and their communications to the following highly compact form, which nonetheless describes the behavior and interactions of all particles and force carriers:

$$L^{SM} = -\frac{1}{4}F_{\mu\nu}F^{\mu\nu} + i\bar{\psi}\gamma^{\mu}D_{\mu}\psi$$

This formula, based on mathematical symmetries, forms the foundation of the Standard Model to this day. The term $-\frac{1}{4}F_{\nu}F^{\nu}$ describes the force carriers. The term $i\bar{\psi}\gamma\, D\, \psi$ describes the particles and their interactions. The whole equation describes which properties determine whether particles detect each other's presence and, if so, how they react to each other. We've seen how it works: three properties and three corresponding forces, each with its own special properties. Even as our measurements have grown more exact, they've remained consistent with the predictions of this theory. But as elegant as the structure of the Standard Model is, and as sturdy as it appears, there was still a problem. A serious one. The Standard Model in this form could not explain why particles have mass.

And what made this problem so serious is that particles *do* have mass—that's a fact we can't deny. On top of that, the mass of the carriers of the weak nuclear force play a crucial role in restricting its operation to a very short range. Reality didn't match up with the theory—which was incredibly frustrating! It looked like you had to choose: *either* symmetry, *or* mass for force carriers and other particles.

Ultimately, three people were able to combine these two worlds: Peter Higgs, François Englert, and Robert Brout. In Belgium, their solution is known as the Brout-Englert-Higgs mechanism. Everywhere else in the world, it's known as the Higgs mechanism. It's a sensitive issue, because even though each of those three physicists contributed to the development of the mechanism that forms the keystone of the Standard Model, it's the name of Peter Higgs that has stuck—to both the mechanism and the corresponding particle.

That brings us to the analogy that best captures the essence of the Higgs mechanism: it's as if space-time is filled with some substance (the Higgs field) to which particles stick. Some particles become stuck very firmly in this field, so they move through space slowly and are heavy. But other particles barely notice the field is there and move through space almost effortlessly. Those are the lightweight particles. This Higgs field would have to be present throughout space, everywhere in the universe—even in empty space, the vacuum. The Higgs field is to elementary particles what water is to the fish in the sea. It's all around them, and some fish can move through it more quickly than others. Although this answers the question of why particles have mass, it still doesn't tell us why particles have the particular masses they do. But we'll worry about that later.

The great advantage of the concept of a Higgs field is that it accounts for why both force carriers and other particles have mass. But it does make the formula describing the motion of all the particles a little less concise. What we end up with is the famous formula for the Standard Model, which you can also find on the coffee mugs and T-shirts for sale in the CERN gift shop:

$$L^{SM} = -\frac{1}{4} F_{\mu\nu} F^{\mu\nu} + i\bar{\psi}\gamma^{\mu} D_{\mu}\psi + \psi_i \lambda_{ij} \psi_j \phi + (D_{\mu}\phi)^2 + V(\phi)$$

Although the formula looks complicated, the addition of the three extra terms by Higgs and his colleagues was fairly straightforward and elegant. As before, the first two terms refer to the force carriers and particles and their interactions. The $\psi_i \lambda_{ij} \psi_j \phi$ term refers to the mass* of the particles, the term $(D \phi)^2$ refers to the mass* of the force carriers, and the final term, $V(\phi)$, describes the Higgs field and particle. The formula contains everything we need, and forty years after the first prediction, we still haven't found a simpler way to account for the fact that particles have mass. But the trick that Higgs used to accomplish this feat was not without cost. Higgs predicted that, if his idea was true, an additional particle had to exist. It *had* to.

To understand most of the consequences of this formula, you'd need a lot of complex math. So you'll have to trust me on a few things—or if you really prefer, you could sacrifice a few months of your life and work them out for yourself. If you've been paying close attention, you noticed the asterisks (*) when I described the mass of the particles (including force carriers). Those two terms tell us that if particles have mass, they also have the potential to annihilate and form a Higgs boson. Those two properties are inextricably linked. That fascinated experimental physicists, because it suggested that you can make your own Higgs bosons in the lab, by using particle accelerators to fire the right particles at each other. Their collision would then produce a Higgs boson.

The search for this particle, which had to be hidden in empty space, required enormous effort to get off the ground. It would take more than forty years before the ingredients were in place: accelerators that could make particles collide with enough energy to create such a heavy particle, and detection equipment so accurate that it could unerringly filter out the evidence of the Higgs boson's existence from the remnants of a particle collision. We will look at these elements in detail. We'll also take a closer look at the people and organizations behind this discovery, especially CERN, the European Laboratory for Particle Physics in Geneva—an amazing lab with fantastic people who together have turned the dream of creating our own Higgs bosons into a reality.

5

Discovery of the Higgs Boson

The mission before us was clear: to discover a particle hidden in empty space. Or at least a particle that *might* be hidden in empty space. The rules of play of the Standard Model and the clues found by Peter Higgs told us that if this Higgs particle existed, we should be able to produce it by making high-energy particles collide. That's a lot of ifs, ands, and buts, and knowing you can do something "in theory" is never a guarantee that you'll succeed in practice. After all, we know how to dig a tunnel under the Atlantic, *in theory*. We can make a full-scale replica of the Great Wall of China out of Legos, *in theory*. And world peace is a real possibility, *in theory*. But in practice, there are often too many obstacles in the way. Our search for the Higgs boson required three things: (1) a particle accelerator powerful enough to produce Higgs bosons, (2) a device that could then detect the Higgs bosons in the wreckage of the collisions, and (3) probably the most difficult part, an intensive fifteen-year-long collaborative effort by thousands of physicists from around the globe. Each of these facets of our mission involved various steps, every one of which looked like an insurmountable obstacle. But for some reason, scientists are often hopelessly optimistic, and to them *in theory* always means *we can do it!*

The steps we needed to take were huge. For example, to produce enough Higgs bosons we needed a particle accelerator almost ten times as powerful as the previous model, the ultra-high-energy Tevatron at the Fermilab particle physics and accelerator laboratory near Chicago. Since we also needed to see what happens in all those collisions, we had to come up with new experimental techniques: everything about the project had to be bigger, faster, more robust, more accurate, and more data-intensive than ever. At the start of the project, the detectors we needed existed only in scientists' dreams—and mostly in their nightmares.

While most people shake their heads and keep inching forward along the same old path, one small group of people dares to look beyond the horizon. They are the kind of people who sometimes try a completely new route, and whose nightmares occasionally turn into dreams come true. In practice, their projects tend to be utter failures, but every now and then there's a brilliant breakthrough, the kind of project that makes everything fall into place and moves science a giant step forward. It's the stubborn, self-confident, open-minded, and sometimes naive attitude of these scientists (and some politicians) that gets these projects off the ground in the first place. Projects like this one, which we knew would take at least fifteen to twenty years from the original concept. The complete success of this adventure was a minor miracle. Everyone feels that way, even the people who kept the project going for years.

Now back to our story. Although technology is a basic necessity in this type of scientific quest, it was not all we needed. There also had to be a research center: a physical location for the particle accelerator, the detectors, and their infrastructure.

A place with areas for meetings and discussions. A permanent home and workplace for several thousand physicists as well as technicians and support staff. The trickiest part of this whole enterprise may have been the sociological aspect. To make it all work, thousands of physicists and technicians from more than a hundred universities all over the world had to collaborate. These are the same scientists that often work in relative isolation at their universities, focusing most of their energy on their own careers and putting their own work in the spotlight. How would they divide the boring jobs among themselves, make sure that good ideas get noticed, and limit the role of national and partisan interests? In short, how would they hold it all together? And on an even more practical note, how would we fund and oversee a project that requires much longer-term and larger-scale financing and planning than garden-variety scientific and political processes?

Looking back, it was not only possible *in theory*—we really did it! We found an international research center with the facilities and infrastructure to house the experiment: the European Laboratory for Particle Physics, CERN, in Geneva. And we managed to push the limits of our technology far enough, and to set up two major experimental collaborations that succeeded in converting industrial technologies into huge detectors: the CMS and ATLAS experiments. Each of these collaborations has involved more than two thousand scientists who have been working together for more than fifteen years, and the two experiments are now engaged in a (friendly) competition to be the first to explore the new world. Despite this rivalry, both groups of scientists work in the same building at CERN, the illustrious Building 40. They share meeting rooms, the cafeteria, sports clubs, and sometimes even beds—there are plenty of "mixed" CMS-ATLAS couples. So it's a healthy, friendly kind of rivalry, but it does energize the scientists and spur them to new heights, like the rivalry between top athletes. No one wants to come in second, especially not if the first prize is immortality. This sociological, technological, and scientific miracle was possible only because a large group of people shared a dream, and we knew that to have any chance of making it come true we had to work together. And our dream did come true. On July 4, 2012, both the ATLAS and the CMS experiments announced that they had found evidence in the collisions at the Large Hadron Collider (LHC) for the existence of the Higgs particle. Our suspicion was true: empty space was not empty, but filled with the Higgs field. We had discovered what gives every particle in the universe its mass.

I would like nothing better than to tell you right away how we concluded from the collisions in the particle accelerator that Higgs bosons really had been created. But because the success of this adventure depended on several remarkable components, it's helpful to take a quick look at each component in turn and learn more about what makes them so remarkable. So let's take a tour of the CERN particle physics lab in Geneva, the particle accelerator itself, and the device that ultimately enabled us to analyze the collisions.

CERN, the European Laboratory for Particle Physics

For more than fifty years now, *the* place for particle physicists to meet, do research, and extend the frontiers of our knowledge has been CERN, the European Laboratory for Particle Physics near Geneva, Switzerland. The institute occupies a few square kilometers in the Pays de Gex, tucked between the city of Geneva and the last foothills of the Jura Mountains.

It's a place with an almost mythical status, where the brightest minds in particle physics—both theorists and experimenters—come together and make all sorts of discoveries, and where you often run into Nobel Prize winners (present and future) in the cafeteria. But of course, it's also simply a science park like so many others. A place where research is done, where every year hundreds of young students are educated, and where we, the community of particle physicists, hold many meetings. There is more to CERN, as an institution, than the research that goes on there. Besides being the Mecca of particle physics, CERN also embodies an idealistic form of collaboration: people from different countries and cultures meet in an atmosphere (almost) free of politics, one in which they can discuss and expand their knowledge together. They engage in a form of collaboration that, despite their many differences, truly works. Why? Because they have a shared dream and the freedom and flexibility to welcome any and all ideas. It's an "ordinary" science park, yet it has more than earned its mythical status, which I certainly won't challenge here. On the contrary, that's one of CERN's strengths.

CERN is an international organization financed by its individual member states. CERN's task is not only to make research facilities available and to play a major role in coordinating particle physics research all over the world, but also to offer a place where new ideas can find an audience and be tested. For the purposes of our story, CERN is responsible for the design, construction, and operation of the Large Hadron Collider, the particle accelerator that causes the high-energy collisions in which Higgs bosons can be produced. Universities and research institutes are given the opportunity to design equipment and use it to study the collisions. The scientists employed by CERN itself, about three thousand in all, mainly play a technical, supporting role, and at any given moment thousands of other scientists from around the world are roaming the few square kilometers of the site. Physicists and technicians constantly pour in and out, visiting for work or meetings. Some stay for a few days, others for a year. In other words, it's the world in miniature: a remarkable sociological experiment, and a tremendously inspiring and challenging environment.

To a particle physicist, CERN is simply a workplace, which besides its experimental facilities also has technical departments and the renowned CERN Computing Centre—where the World Wide Web was invented—as well as two hostels, cafeterias, coffee corners, and a bank. The site is covered with buildings whose numbering system confounds even Nobel Prize winners—though you gradually develop a feel for it during your stay at CERN. For example, I've had offices in buildings 13, 16, and 27, and I'm now in building 40. It's a very dynamic environment, especially stimulating to a young researcher, because you have all the world's great names, knowledge, and expertise packed into a few square kilometers. An incredible luxury, and we always try to impress that on doctoral students who will be staying there a while: soak up all the expertise *and* make your own contribution. Alongside the science, it's also home to hundreds of young people from around the globe, all bursting with energy. Ideal conditions for sharing a meal, going out on the town, or taking a weekend drive through the Mont Blanc tunnel to Italy for an A.C. Milan game. Not too often, of course, because at CERN there's not really a difference between weekends and ordinary working days, and in busy periods not even between day and night. That's not just because scientists at CERN work all the time, but also because you're collaborating with people at universities and research centers all over the world.

As for me, my very first visit to CERN quickly won me over. I had chosen to study physics in Utrecht, drawn by the magic of the theory of relativity, quantum mechanics, and the mathematics I'd read about in books and newspaper articles. But the first year of coursework was followed by a series of subjects that were actually pretty dull, and it wasn't until the final year of my studies that physics began to fascinate me again. Particle physics, an elective subject, not only had a fantastic teacher, Adriaan Buijs, but also was a field that brought together all my interests. Marvelous! After a quick visit to CERN with friends, I followed the path that so many others had followed before me, gradually falling under the spell of particle physics. I wrote my master's thesis on a CERN research project, applied for a spot as a CERN summer student, and was selected. It was a life-changing experience, one that expanded my horizons in one fell swoop from Utrecht and the Netherlands to the whole world. You see, the CERN summer student program selects a few students from each member state to spend three months in Geneva, take classes from the giants in the field, and even participate in a real research group.

Suddenly I was working in a group with a cheerful Norwegian leader and an Italian technician, and at classes and lunch I was sitting next to students from Greece, Finland, Germany, and Portugal. Students just like me, so crazy about this one little corner of physics that our cultural and linguistic differences melted away. On top of that, they were the kind of people who join you for a beer after work. All this makes CERN so much more than a particle physics institute. It's truly a place where scientists from many countries meet and share ideas. To me, CERN is the embodiment of the limitless possibilities open to us when the world bands together to work toward a shared dream. It radiates not only a passion for physics but also a sense of human connection. Of course, we have our share of difficulties even in that ivory tower, but we stand together, sharing both our disappointments and our triumphs with the outside world. I am forever proud to be part of that.

What makes a joint research platform like CERN so valuable, then, is not simply that it can tap into so much more funding than a national institution, but also that it provides a Europe-wide—no, more than that, a *worldwide* research community and strategy.

The reason we keep making our particle accelerators more and more powerful is not only to make our probing fingers (that is, the projectiles) ever tinier, so that we can explore smaller and smaller structures. We've also seen that we can convert the kinetic energy of the colliding particles (that is, the energy of their movement) into new matter: that is, new particles. And if you manage to go on increasing the energy of the collision, there comes a point when the energy is high enough to make not only all the familiar particles, but also the illustrious Higgs boson. That's why CERN's main job was to design a new particle accelerator that would cause lots and lots of collisions with enough energy to produce Higgs bosons. If they existed, of course.

The accelerator consists of two thin tubes in a tunnel seventeen miles (twenty-seven kilometers) long in which protons are accelerated to an energy of six and a half tera electron volts (TeV) and then fired at each other. That adds up to thirteen TeV of energy available in the collision itself. Although the energy of these protons is tiny to human standards, roughly equivalent to that of a flying mosquito, it is enormous for a single elementary particle. It is the highest energy we can give an elementary particle on earth today, and with the enormous number of protons whizzing around in the LHC, the total energy stored in the machine is equivalent to that of a speeding train. The tunnel containing the accelerator is a few meters in diameter, looks a little like a subway tunnel, and is about three hundred feet (one hundred meters) underground, well protected in a stable geological formation. An underground location is ideal for this type of machine: the temperature is constant there, any radiation will not put anyone at risk, and the beautiful landscape remains unaffected. The tunnel was not in fact built for the LHC, but had been there since the 1980s, when it was dug to accommodate the Large Electron-Positron Collider (LEP). Now, I don't have a bad word to say about the old LEP—not only because it was a fantastic machine that taught us all kinds of things, but also because my doctoral research was based on data collected there. But to reach high energies, we needed a new machine, one that would shoot protons at each other instead of shooting lighter electrons and antielectrons at lower energies: the Large Hadron Collider.

Although over the past forty years, different types of particles have been fired at each other, the technique for speeding them up has stayed almost the same. Particles in a vacuum tube are given a shove with an electromagnetic field and then led around in a circle with magnets, after which they receive another push. With each circuit, the energy of the particles increases, and we have to adjust the power of the magnets to make sure that the particles, which now have higher energy, won't fly off the curved path. It's a fairly delicate procedure, but the collider operators at CERN have been doing it for years. The particles accelerated in the LHC are protons, and it takes about twenty minutes to raise them to their maximum energy and then fire them at each other.

But why protons, and why a maximum collision energy of thirteen TeV? Making electrons and positrons collide, as we have done in the LHC's predecessor, the LEP accelerator, has offered an enormous advantage as an experimental technique: the collisions produce a relatively small number of particles. That makes the results easy to interpret in terms of underlying processes. "Why change a winning team?" you might think, but in this case, it's a little more complicated. Because electrons and positrons are so light, they lose a lot of energy as radiation with each circuit around the accelerator. That's dictated by the unbreakable laws of nature. And with hundreds of billions of electrons making more than ten thousand circuits a second, the LEP needs more than twenty megawatts of power to keep the particle beams at a constant energy. This enormous amount of energy would have to increase exponentially as we aimed for still higher energy levels. A practical impossibility, in other words.

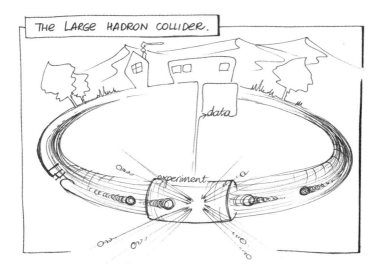

By smashing together protons, which are much heavier, we can completely avoid the problem of losing energy to radiation. Instead, in our pursuit of the most energetic collisions possible, we run up against a new obstacle: the maximum strength of the magnets needed to lead the protons around the ring and ensure that they complete the full circle. With a larger tunnel than the accelerator in Chicago and more powerful magnets, the LHC would be almost ten times stronger than the most powerful particle accelerator before it. That turned out to be enough.

In the Large Hadron Collider, protons don't spin around independently, but in bunches, or clouds. There are several thousand such clouds, each of which consists of around a hundred billion protons. The LHC is not a stand-alone machine, but part of a large network of particle accelerators. It's like riding a bicycle: you start in a low gear, and when you can't pedal any faster, you shift into a higher one, finally reaching top speed in the highest gear. Likewise, when we accelerate particles at CERN, the protons that will eventually collide in the LHC are first raised to ever higher energies in a series of smaller accelerators. Each of these accelerators was once the CERN flagship, and it's great to see the old infrastructure still in daily use. In our current research, we still stand on the shoulders of giants from the 1950s, 1960s, 1970s, 1980s, and 1990s—literally every day. The first time that protons were about to be fired into the LHC tunnel, we were all holding our breath, waiting to see if we could really steer them in circles around the machine. We soon succeeded, and euphoria swept over all the people in the control room and in labs around the world. The machine worked! The next step: focus the beams and lead them through the imaginary needle's eye so that they would collide.

A proton collision doesn't result in such a perfect, straightforward picture as a collision between electrons and antielectrons. The thing is, even though protons allow us to achieve much higher energies, they also have a downside. Here's the problem: protons are made of quarks and gluons, so their collisions lead to large numbers of particles in the detector. That makes it hard to tell exactly what's going on. This is why many people compare proton-proton collisions to smashing together two garbage bags, or two alarm clocks, so hard that they both break open. Interpreting the resulting mess correctly, in terms of fundamental processes and constants of nature, is astonishingly complicated. In fact, almost everyone who sees an image of this type of collision for the first time calls it an impossible task. But at the same time, it's an exciting challenge for physicists. After all, it's possible . . . in theory!

Despite what many visualizations suggest, the LHC does not form a perfect circle. Most of the machine does make a neat circle, and those parts of the path are filled with the enormous magnets that guide the particles back to the part where the acceleration takes place. But there are also a few straight sections. In one of them, the particles are accelerated, and in the others, the beams can be crossed so that the protons collide. To make it much more likely that this will happen, we focus the beams just before they reach the collision point, the same way you can concentrate a beam of sunlight on a single point with a magnifying glass. Since we're effectively sending each of the two beams through the eye of a needle, the protons are almost certain to collide. Those are the points at which new particles are created, and no one will be surprised to hear that we have dug large underground chambers for our detectors in exactly those spots, so that we can observe the post-collision debris. The Large Hadron Collider's specs must have been pure science fiction back when it was designed. The numbers involved can still make your mind reel.

MAGNETS

At eight Tesla, the magnetic field in the LHC is very impressive. That may not mean much to you, but it's an incredibly strong field, especially considering the fifteen-meter length of LHC dipole magnets. A magnetic field is generated when electric current passes through a coil of wire, and it forms in the middle of the coil. This basic principle is fairly simple: the stronger the current, the stronger the magnetic field. But to produce a magnetic field strong enough for the LHC, we have to do more than just roll a long copper wire around the beam pipe and send a current through it. The amount of current required to generate the magnetic field we need would give off so much heat that the wire would melt instantly. So that's a no-go. That is, unless you make the wire out of a special material and dip it in liquid helium to make it superconducting. That prevents energy loss when a current passes through the coil (in other words, the coil won't heat up) and gives us a very large current that we can use to generate a powerful magnetic field. This "trick" is performed in almost every lab in the world, but to use it on all 1,200 fifteen-meter-long LHC magnets, we would need to reserve a significant portion of the global supply of helium for use at CERN. So we've gone ahead and done just that, and the LHC has now earned the title of the world's largest refrigerator.

If there's an accidental short circuit in one of the magnets, then not only does the electricity have to be diverted quickly (to prevent the magnet from melting), but the other magnets in the ring also have to temporarily generate a slightly stronger magnetic field, so that the protons can complete their circuit as planned. It was this standard procedure that went horribly wrong after the first brief period of successful collisions. The power was not rerouted in time from one of the magnets, so the helium heated up, and because a design flaw prevented it from escaping, it went searching for its own way out. The forces unleashed were phenomenal, and over a several-hundred-meter stretch, the magnets were wrenched completely loose, even though they'd been anchored securely in solid rock. A disaster. But fortunately, the cause soon became clear. Despite all the safety measures in place, these things can happen to our one-of-a-kind machines with specifications that push the limits. The LHC is a very different kind of beast from the cars or televisions that come off the conveyor belt by the millions.

Luckily, we managed to heat up the magnets, open up the machine, and repair it fairly quickly, but another challenge awaited us. After the repairs, the magnets had to be cooled down again. And unlike freezing a bottle of water in your kitchen freezer, which takes just a few hours, freezing the entire magnet section of the LHC to a few degrees above absolute zero took a few months. We eventually decided to repair the ring in stages. As a result, we could use the magnets only at half power for the first two years. That halved the energy of the collisions but also allowed us to collect collision data in a safe and stable way for a long time. And it was those collisions, with a reduced collision energy of seven to eight TeV, that gave us the proof that the Higgs boson exists. So in hindsight, we made a good choice.

A BEAM OF PARTICLES

A single proton flying around the LHC at maximum velocity has the same energy as a mosquito in flight. That may not sound so impressive, until you realize that it's not isolated protons zooming around the LHC, but thousands of clouds of about a hundred billion protons each. And although the energy of a single proton may not amount to much, the energy of all those protons put together equals that of a train at full speed. That train, which makes a complete loop ten thousand times a second, has to be steered through the tunnel with great delicacy. Make just one little mistake and the beam could slip out of control, shooting straight through your experiment and doing irreparable damage. So the beam is tracked on its route around the machine with great precision, and after fifty years of experience with particle accelerators, we can rely on CERN's LHC beam operators to do an expert job.

To understand just how precise this tracking is, consider the ease with which the operators compensate for the tiny changes in the tunnel's circumference that occur naturally in the course of the year. These changes include monthly variations in the length of the tunnel caused by tidal effects in the earth's crust (like ocean tides, but in the earth itself), which make the tunnel expand and contract slightly. The differences are mere fractions of a millimeter, but they're clearly visible as the beams make their rounds. Another variation is linked to the water level in the deep Lake Geneva. As the weight of the water distorts the surrounding crust of the earth, it distorts the tunnel a little as well. Both of these tiny differences can be detected and corrected effortlessly by the LHC beam operators.

As the proton beams race around the ring, they are about the size of the dot made by those laser pointers used in presentations. We sometimes ask people to take a two-euro coin and look at the side with the map of Europe. If you find Spain, that's about the size of the beam in the tunnel at full energy.

But when the beams get close to a collision point, they are so highly compressed that the spot where they meet—the proverbial needle's eye—is smaller in width than a human hair. It's pretty hard to aim so accurately, but on the other hand, if the beams miss each other, they just make another loop and meet again in the same spot 0.0001 seconds later. For the people who steer the beams, that's more than enough time to bend them slightly, using magnets, so that they *will* collide the next time they cross.

The chance of producing a Higgs boson in a single proton-proton collision can be accurately calculated with the help of the Standard Model theory. That chance is extremely slim. That's not so strange, because lightweight particles are much easier to make than heavy ones. Anyway, if the Higgs boson were easy to make, we would have seen one a lot sooner. We can't do anything about how slim that chance is, but there's nothing to stop us from causing a mind-blowing number of collisions and then picking out only the interesting ones. So that's our strategy.

The LHC is designed so that the distance between the two proton clouds (in other words, the beams) is about seven and half meters. Because the beams move at nearly the speed of light, they cross forty million times a second at the interaction point. If you know that about ten to twenty protons collide in each of these crossings, then you also know that nearly a billion collisions take place each second. Like the other heavy particles, the Higgs boson is extremely short-lived, and you have to reconstruct the collision from the debris that reaches the detector, as if you were putting together a puzzle. And you have to work fast, because twenty-five billionths of a second later, it's time for the next collisions. That puzzle-solving is a job for the experimenters, and they can handle it. But again, it takes a variety of technological and computer wizardry to get it done.

The ATLAS Experiment

In each of the four places where the proton beams are made to collide, measuring devices have been built, in major partnerships between universities around the world, to study the collisions. The two biggest experiments, ATLAS and CMS, were developed to detect many different particles in collisions, but the discovery of the Higgs boson was the absolute top priority.

Particles produced in the heart of the detector shoot outward and pass through the different detector layers that are part of every traditional particle physics experiment. Like the accelerator itself, the detectors have much better specifications than in earlier experiments; they're more precise, more accurate, faster, more robust, able to withstand high doses of radiation, and capable of measuring those high energies more accurately. If you look at the size and complexity of the detectors and software needed to analyze the collisions, it's not so strange that thousands of physicists are involved. The numbers are overwhelming. The ATLAS detector is a cylinder some 150 feet (45 meters) long and 80 feet (25 meters) high, about the size of the White House, crammed with high-tech electronics and sensitive equipment. The detector seals off the collision point as hermetically as possible, so that no particles can escape unnoticed. A completely hermetic seal is impossible, because the proton beams must have a way to enter the detector from both sides, and there has to be enough room for electricity and coolant to go *into* the system and for signals and heat to come *out*. Coming up with the optimal design was a huge puzzle, especially because we didn't know the exact dimensions in advance and weren't even certain whether all the detectors that make up ATLAS would work.

The fact that an international scientific experiment like this one has survived and succeeded is a minor miracle. While big companies often have strict hierarchies so that they can operate efficiently, in this case an organization of two thousand highly educated people had to be made up of highly autonomous groups with ten to twenty members. A very complex proposition, because how can you make sure that Japanese doctoral students will work together with German professors or collaborate "constructively" with a very competitive research group from the United States? It makes the building of the Tower of Babel look like child's play. And how do you assign all the tasks? Sure, you'll find plenty of volunteers for the final presentation at the big conference, but who will sit in a control room at 2:30 a.m. on a Sunday morning just in case something goes wrong with the detector? And who will work on one of the many small, basic components like calibrations or computer simulations? Jobs like that may not be so sexy, but they are essential to the end result. Fortunately, there's a good system for making sure that the work is divided fairly among the scientists. Although it may sometimes seem inefficient—and to be honest, it often *is*—the flexibility of this approach leaves room for unconventional ideas, and for listening to them with an open mind. A good idea is a good idea.

When the LHC runs according to specifications, there are about a billion collisions a second at the interaction point. But most of them are completely uninteresting. The protons merely graze each other, or they produce lightweight particles of long, familiar kinds. To really enter the new world, the protons have to crash into each other head on. And even then, the Standard Model predicts that most collisions will produce nothing more than a few quarks or gluons. The truly interesting collisions are those in which heavy particles are created: Z bosons, top quarks, or one of our dream particles that no one has yet discovered—like the Higgs boson until 2012. It's a simple fact that we'll never be able to study all the collisions in detail. Collecting data from a detector takes much longer than the twenty-five nanoseconds that pass before the next proton clouds cross in the heart of the detector. And even if we could gather all the measurement data in that time, we don't have enough disk space for it all. The information about all the particles in a single collision takes up about two megabytes, approximately the same amount of space as a photo on a cell phone. That may not sound like "big data," but if you take a thousand photos a second for a few months straight, the total volume grows to a substantial number of petabytes. That's big data for sure: a petabyte is a thousand terabytes, or several thousand of the back-up hard disks that many people have at home. How can we select the thousand interesting collisions each second, and how well can we identify all the particles produced?

The experimental techniques used in the big detectors to identify particles are the same ones we saw in earlier chapters. The velocity and charge of electrically charged particles is measured by looking at the paths the particles follow in the heart of the detector. We recognize those paths because the charged particles leave a trail of signals (small electrical currents) as they pass through the thin detector layers. Then we estimate their energy by destructive measurement, bringing the particles to a complete stop. By combining the information from the different detector elements, we can tell apart the different particles, much the same way we can tell apart the very different tracks left in the snow by a human being and a rabbit. That leaves just two types of particles we're at risk of missing, because they pass through the calorimeter almost unaffected: muons and neutrinos. Muons are electrically charged, so they can be identified by adding a couple of those thin detector layers right on the outside of the detector. Neutrinos remain invisible to ordinary particle detectors, leaving the detector unobserved.

The particles we detect are the pieces of a puzzle that offers us insight into the new world, so it's crucial not only to identify the different types and measure their properties, but also to know exactly *how* accurate our measurements are. Are we wrong one out of every hundred times we say a particle is a muon, or only one out of every hundred thousand times? And if we measure a particle's energy, is the margin of error 0.5 percent or 20 percent, and how do we find out? Problems with obtaining these figures and proposed strategies are the subject of many meetings. Next I want to give you some idea of the incredible precision of the detectors, and of how we gained experience with our new equipment, both before and during our observations of collisions.

To use a new measurement device to take photos of a new world, you have to really get to know the device inside and out. Otherwise, when you finally see something strange, how will you know it's a new observation and not just a flaw in your equipment? For example, if a National Geographic photographer sends black-and-white photographs from a first visit to a previously unknown island, no one will wonder whether or not there are colors in that part of the world. And if he then sends color photos of pink giraffes eating pink bananas from a tree, we'll understand that there's something wrong with his camera's color settings, since we all know that giraffes and bananas are yellow. But it's not that easy when the photos are coming from Mars or the bottom of the ocean. Without any frame of reference, how can we tell whether the photos we're seeing are accurate? How can we estimate the size of mountains on Mars without anything familiar to compare them to, or figure out the color and size of a new deep sea creature? This shows that it's vital to get familiar with your equipment, whether you're studying a new landscape or searching for new phenomena in particle physics.

Fortunately, we can build up confidence in our equipment over time, because even in the new world of new collisions, some very familiar types of particles are created. Like the giraffe in the color photo, these particles have such well-understood properties that you can use them to calibrate the detectors. You can gradually increase the complexity of these standard signals until finally you are able, with confidence, to identify strange phenomena as *truly* strange.

To get a sense of what makes the ATLAS detector so special and what's involved in calibrating it, it's interesting to look at a couple of examples. Though I'm presenting the equipment here as if it's the state of the art, it's not really. You see, most of its components date from around twenty years ago, and when it comes to technology, what was science fiction twenty years ago is now commonplace—at least at CERN. The fact that this equipment is still seen as cutting edge makes it all the more impressive.

THE PIXEL DETECTOR: A CAMERA ON STEROIDS

The heart of the ATLAS experiment is a detector consisting of three thin layers with a total of eight million very small (0.05 × 0.4 mm) detection elements, known as pixels, which register a small current when an electrically charged particle flies through them. Those layers form a cylinder that is placed as tightly around the beam pipe as possible so that the exact collision point can be determined. You might reasonably compare it to an eighty-megapixel digital camera, even though they don't make retail cameras like that yet. And if you want one anyway and go to the electronics store, or to Canon or Sony, to ask when they'll be available, they'll tell you the technology doesn't exist yet and probably won't hit the consumer market for another few years. If you add that you'd also like the camera to be able to take around a billion photos a second, you'll get a funny look—in fact, they'll probably laugh in your face and walk away.

Yet that was the equipment, and those were the specifications, devised twenty years ago (!) so that we could advance into the new world. You might conclude from a conversation like that at a camera store that it's impossible, and we just have to wait for developments in the industry—end of story. But it doesn't necessarily have to end there. If you can't wait for the industry to sell you a camera, or don't want to, the implication is clear: you'll have to invent your own. That's unbelievably complicated, but if you can recruit the world's smartest physicists and engineers, and if you want badly enough to make your dream come true, then you can do it. More than that, it's been done! A device that meets this description already exists and is now one hundred meters underground, at the heart of one of the biggest experiments in the world, the ATLAS experiment.

THE CALORIMETERS: ENDLESS CALIBRATIONS

We measure particles' energies by bringing them to a complete stop, and we call the detector layer where that happens the calorimeter. The principle is fairly simple: the deeper a particle penetrates into the detector, the higher the energy when it entered. In the case of electrically charged particles, we get a fair impression of their energies from their deflection in the innermost part of the detector. But for photons (particles of light) and other uncharged particles, the information from the calorimeter is all we have to go on. Measuring the energies of photons is especially important, because the Higgs boson can decay into two photons, and that's actually one of the most important signatures for us to focus on in our search.

Calorimeters have been used in experiments for decades, so we have a pretty good idea how to design a detector with whatever features or level of accuracy we need. But "pretty good" is not good enough, and to figure out exactly what to do in this case, we need to know the exact effects of different designs and the influence of temperature and other factors that could interfere with measuring the energy. That means testing, testing, and more testing. We could do that during the experiment, but if problems arise at that stage, it'll be too late to fix them, so we prefer to test in advance and under controlled conditions.

So how does that work? Well, to test whether your new thermometer shows the right temperature, you can dunk it in ice water and then in boiling water. It should show temperatures of 0 and 100 degrees Celsius, respectively. Those freezing and boiling points serve as your benchmarks. For example, if your thermometer reads −2 and 98 degrees, then you know that it's consistently two degrees off. That's no problem, of course, because when you use the same thermometer for future measurements you can easily correct for that slight offset. We follow the same procedure with particle detectors. At CERN, we can test new ideas for experiments or calibrate existing detectors with the help of what we call test beam facilities. These produce beams of a particular type of particle that has an exact known energy. By placing the detector in the path of these beams, you can study how your device reacts and how well it can identify the particles and their properties. These test beam programs are where the real hands-on work of particle physics is done, and I couldn't tell you how many nights I've spent at CERN testing the electromagnetic calorimeter. It's serious fun, and seriously educational, but also seriously exhausting and often incredibly frustrating. Still, when you see how accurately you can measure the energies of electrons and photons in the CMS experiment, and the high quality of the results, you know it was all worthwhile.

MUON DETECTORS: MONITORING SMALL SHIFTS

From the calorimeters, we go on to the outermost shell of the experiment: the muon detectors (muon chambers). As I'll later explain in more detail, detecting and measuring muon particles is one of the main keys that opened the way to discovery of the Higgs boson. One of the clearest identifying characteristics of the Higgs boson is its decay into four muon particles. These muon particles resemble their lighter brother, the electron, in a couple of important ways. As a muon passes through the layers of the detector, it leaves signals, and the deflection of the muon gives information about its velocity and electrical charge. In that respect, it's a lot like any other charged particle, but thanks to one of its unique properties, it can pass almost effortlessly through the thick plates of iron in the calorimeter that stop almost all other particles. So by adding a couple of sensitive layers that reveal electrically charged particles, like those in the heart of the detector but all the way on the outside, we can identify muons for what they are.

Classifying a particle as a muon is an important step, but not yet enough. If we want to know whether two muons have the same parent, we have to be able to associate them with each other, and to do that, we need to know their exact properties. We know that a Z boson can decay into a positively and a negatively charged muon, so when we detect that combination, we can reconstruct that a Z boson must have been produced. And since the properties of the Z boson are very well known, we can then use those muons to study and calibrate the detector's response. In principle, this is just like the example we saw of calibrating the thermometer. Let's keep this fact in mind as we turn to collisions in which four muons are produced and we have a chance of discovering the decay of a Higgs boson.

One interesting problem arises in cases where a muon travels with such extreme velocity that it's deflected by only a fraction of a millimeter as it passes through the huge detector. To observe that tiny deflection and use it to figure out the muon's velocity and electrical charge, we need to know the exact position of each detector. If we're even a millimeter off, we may confuse a positively charged particle with a negatively charged particle and end up with a totally wrong impression. Then we might classify the collision as uninteresting, even though it may have involved a Higgs boson. So it's important to know exactly where the detectors are and to keep close track of their position throughout the period when we're observing collisions.

Right after we installed the muon chambers deep underground on the outer perimeter of the ATLAS detector, the year before the first collisions, we were lucky enough to receive a little help from the cosmos. We saw earlier that cosmic rays from outer space produce muons when they hit the atmosphere. These muons not only easily reach the earth's surface, but also pass straight through the soil and rock over the LHC like a warm knife through butter. So they occasionally pass through ATLAS's muon detectors, even though these detectors are more than 150 feet underground. Long before the start of our colliding beams, that allowed us to obtain a clear idea of the exact position of the muon chambers. But because the muon detectors move and change shape when the temperature changes and whenever the magnet is turned on or off, we also had to develop a monitoring system that keeps track of the shifting position of those immense muon chambers with very great precision. The brain behind this monitoring system— RASNIK, developed at Nikhef in Amsterdam—is Harry van der Graaf. His technique for monitoring deformations right down to fractions of millimeters over distances of dozens of meters is now also used in major infrastructural projects like bridges and viaducts.

The many components of the ATLAS detector form a highly complex machine whose numerous detector layers enable it to identify particles and their properties. If we pool all the information it gives us, we have everything we need to track down the heavy particles that decayed into the particles we measured. So what particles *did* we measure in the collisions, and did they match the predictions of our theory? Answering that question is the only way to learn anything about the theory, and therefore the only way to tell whether or not we need a Higgs boson to account for all the collisions we observed.

THE ATLAS CONTROL ROOM

All year round, every part of the ATLAS experiment is monitored day and night. So hundreds of meters above the experiment itself, there are always a few physicists (the "shift-ers") on duty in the ATLAS control room. With its hundreds of computer screens and a half dozen semicircular stations, it looks a lot like the bridge of a spaceship, or a NASA or SpaceX control room. Each member of this small group of physicists, known collectively as the shift crew, is responsible for a particular aspect of the experiment. If a problem arises that can't be solved automatically, or dealt with quickly by the shifter, they call in the expert, who goes straight to work on it. Even if it's 2:00 a.m. This system requires every ATLAS physicist to sacrifice at least one month a year for the collective good. You can make your contribution by monitoring the detector and keeping it running, or you can help with software mainte-nance or one of the many calibration tests. We really are all in it together, and our cooperative approach is not a mere luxury, but an absolute necessity for the success of our venture.

Even though most scientists don't have much patience with formal organizational structures, a cooperative effort by thou-sands of scientists does require some ground rules. We have no choice. There's a lot of money at stake, and we need to know who will make the big decisions. What technology do we use? How do we analyze the results? Who will be in charge, and what powers will that person have? When CERN outsources work to private companies, it has to distribute the jobs care-fully among the participating countries. With experiments, too, balance is key. We all want the best people for every job, and that requires recruiting a diverse group of physicists with a variety of perspectives. When assigning conference presen-tations and important positions like group leader, we also con-sider how much credit each physicist has earned by doing the less attractive work. To solve this complex puzzle, we've set up a virtual organization that runs parallel to national and university structures. As a result, a more junior scientist could realistically end up coordinating one line of research in the same experiment on which his or her boss is working. Although this system doesn't rule out all forms of office politics (scientists are sometimes all too human), and CERN has its share of people with a knack for advancing their own careers, it generally works pretty well. Again, it's our shared dream that keeps things running smoothly.

The ATLAS experiment has a number of divisions, and it's not unusual for scientists to contribute to one or two of them. Their activities are highly diverse but all interconnected. There are groups that focus on quality control of one detector component, others that work on computer simulations, and still others that come up with new ideas for detectors, interpret the data to find the fundamental constants of nature in the theory, or pool the research results. For example, the ATLAS Higgs working group is divided into subgroups, each of which tries to measure the possible decays of the Higgs boson with great precision. These subgroups share all sorts of information, from statistics to computer simulations, analysis techniques, and data visualizations. But there are also differences between them. The subgroup studying the decay of the Higgs boson into two photons wants to know exactly how well the electromagnetic calorimeter can measure the energy of photons, while the group looking for decays into four muons is interested only in the latest developments in muon reconstruction. So those scientists need to contribute to, and work with, the muon reconstruction group, and so forth. The result is a tightly woven web of groups and collaborative projects that cover the whole range of topics. The various tasks are sometimes fun, challenging, and groundbreaking, sometimes terribly boring and routine—but all necessary, and distributed carefully among the scientists.

The work of two different groups may overlap—often on purpose. Most analyses and measurements are carried out by a handful of groups operating independently. That may seem very inefficient, and in a sense it is, but by working in parallel, we avoid mistakes and make sure the work is done properly. Besides, different groups may come up with different approaches to the same problem. In the end, of course, only one set of results can be published, and the efforts of each group to make sure its results are chosen lead to heated debates—it truly is the survival of the fittest. This is where the real fireworks happen. Often behind closed doors. But that's understandable: after all, having your analysis, your illustrations, and your figures included in a publication is a cut above a one- or two-sentence mention of your "alternative analysis that produced a consistent result and hence confirms the main result."

Hunting for the Higgs Boson

One of the disadvantages of using protons as colliding particles is that they're not elementary particles but made up of quarks held together by gluons. So in a collision, a quark or gluon in one of the protons bumps into a building block of the other proton. What happens in that reaction is specified by the rules of the Standard Model, which tell you exactly how often the protons will brush past each other and how often they'll hit head-on, creating new matter. But have there been any collisions that can be explained only by assuming that a Higgs boson was produced? And how certain are we of that, really?

The theory tells us that lightweight particles are much easier to make than heavy ones like a Z boson or a Higgs

boson, and that the chance of producing a Higgs is exception-
ally small. To make things worse, even if the Higgs boson exists in the first place, it falls apart after a short but very happy
life. We describe that transition from a heavy particle to lighter
particles as the *decay* of the particle. And just like paleontologists reconstructing dinosaurs and their characteristics from
their bones, particle physicists try to infer from the "stable"
particles (the particles that survive long enough for us to observe them with our detectors) what happened in the initial
collision. In the case of the Higgs boson, we find that most of
the ways it can decay look exactly like the results of collisions
in which very familiar particles are made. And those other
types of collisions happen much more often, so if you make
even the slightest error in identifying the particles, you might
easily draw the mistaken conclusion that a Higgs boson was
produced. And a false claim that you observed a Higgs boson,
which you later have to admit was a mistake, will bring you
not eternal fame, but eternal disgrace.

Before you can say for certain whether or not the Higgs
boson exists, you need to find a group of collisions for which
almost the only possible source is a Higgs boson decay—and
then carefully puzzle over the data to make sure you didn't do
anything wrong. Fortunately, there *are* patterns of particle decay (also called decay channels) that meet this description, but
it's clear that the Higgs boson won't let itself be captured easily.

Our prediction was that the Higgs boson, if it exists, lasts less
than a millionth of a millionth of a billionth of a second and
then falls apart. Its decay can produce a variety of particles: the
exact combination depends on the Higgs boson's mass, which
was a mystery until the discovery. Fortunately, we had figured
out in advance that we would be able to observe different
characteristic decays for each hypothetical mass of the Higgs

boson. So our strategy for discovering it was to focus on two almost unique and very clearly identifiable types of decay:

1. Into two photons (particles of light), or
2. Into four electrically charged leptons (muons or electrons).

Even though these two final states together represent less than 1 percent of all Higgs decays, they are easy to spot among the billions of collisions in the detector. And since their occurrence is expected to be very rare if there is no Higgs boson, they provide the ideal basis for saying with great certainty whether or not the Higgs exists. To pick out these particular decays, we use the strategy described earlier, of searching for "peaks" in our data. Now that we have a focus and a strategy, the remaining steps are fairly clear:

1. **Pick out collisions with two photons or four muons** Not only are these combinations of particles very rare if no Higgs boson was produced in the collision, but our detectors are also extremely sensitive to photons and muons. We can both clearly identify them among the hundreds of particles produced in a collision and determine their properties with great precision.
2. **Calculate the invariant mass of each selected collision** If the two photons or four muons really came from a Higgs boson, they are linked to each other for life. And just as the bones of a certain species of dinosaur always fit together perfectly and produce the same kind of skeleton, the properties of the particles produced in a col-

lision enable us to calculate the *invariant mass* of the mother particle. In this case, the Higgs boson. And when these combinations of particles *don't* come from a Higgs boson, the distribution of the invariant mass shows no clear structure and can be predicted fairly accurately.

3. **Search for a deviation (a peak) in the graph of the invariant mass** If the Higgs boson exists, then one particular part of the graph—the part around the Higgs boson mass—will show a clear deviation from the prediction, because of all the extra collisions there that *do* involve Higgs bosons. The procedure seems so nice and neat when I describe it this way, as a recipe. Yet in practice, it's disappointingly messy. Measuring and identifying the different particles by connecting signals from different detectors, selecting the right collisions, and interpreting them— this was a big job that took hundreds of particle physicists more than ten years to complete.

In the first two years of the LHC, 2011 and 2012, the accelerator produced more and more proton-proton collisions. We looked at collisions involving many different combinations of particles. But as I've explained, the focus was on collisions producing two photons or four muons. This came down to analyzing the graph of the invariant mass to see whether we could find a deviation that pointed to the existence of the Higgs boson.

There are always fluctuations in nature, the well-known distribution around the mean. For example, the temperature of your shower is never exactly the same as it was yesterday, and the number of kittens in a litter varies, too. A deviation from the mean doesn't necessarily signal anything special. Only when that deviation is unnaturally large do we call it an anomaly. We physicists have agreed we won't claim to have seen anything strange unless the deviation is so rare that it exceeds the "five-sigma" threshold. In layperson's terms, we call a deviation (an excess) in the data a discovery when we know that it would only occur once every few million times in experiments like ours, *if* nature worked according to the old, established theory. In our case, a theory without the Higgs boson.

So every time we see a deviation in a graph, our first question is always, "What is the probability that this type of deviation would be produced in a world without the Higgs boson?" The answer to that question indicates the probability that it's a coincidental fluctuation, and tells us whether or not we're looking at something new. Fortunately, we know just how to answer this question, even if it does lead us into the world of statistics. (Don't fret, you've already conquered the hardest equations of particle physics, so you can also handle this.) I could go ahead and tell you the numbers, but first I'd like to present an analogy that may give you a better understanding of this problem and how we approach it.

The problem when we look at collisions, especially the distribution of the invariant mass, is that we don't know *whether* a Higgs boson is hidden in the data at all, and on top of that, we don't know *where* it will pop up if it does exist. To show you how we tackle this problem, I want to offer you a little analogy. Imagine a bag full of dice—twenty, to be exact, one of which may be a trick die. And when I say a trick die, I mean that each side has the same number of dots. For example, each side of the die may have two dots. Or each side may have four.

Your job is to find out whether there's a trick die in the bag. That sounds easy, until I tell you there's another restriction: you're not allowed to look inside the bag. All you can do is ask a friend to take out one die, roll it, and then tell you the number of dots before putting it back. But you can repeat that process as often as you like.

You can see that as long as there's no trick die in the bag, each number of dots will come up about equally often. And if there is a trick die, you'll eventually notice that a certain number of dots comes up more often than the others. But how will you reach that final conclusion, and when do you know for sure? After just a few rolls, you may have very different results for different numbers of dots. After rolling a die ten times, for example, your friend may have rolled four fours and no sixes. But at that stage, it would seem a bold claim to say there's a trick die in the game and that all of its sides have four dots. It's still possible there's no trick die at all. This distribution could be pure chance. In fact, there might even be a trick die with six dots on each side, since the probability of pulling out the trick die (if there is one) is only one in twenty. Of course, after a hundred thousand rolls it's very easy to reach the right conclusion—but maybe you can see the problem. Each new roll gives you more information about whether or not there's a trick die in the bag. But at what point do you feel (or know) you can say with certainty that there's a trick die or that they're all fair dice? Take your time to think about the problem and what you would do to reach a conclusion.

Suppose you've rolled sixty times and have an excess of fours. After sixty throws, you'd expect four to be the result ten times on average, just as you'd expect one, two, three, five, and six to come up an average of ten times each. But let's say you've rolled a four not ten but sixteen times. The question we can ask then is whether this result could be pure chance, or whether it can be explained only by assuming that there really is a trick die in the bag with four dots on each side. To answer this question properly, we need to calculate the probability of rolling a four at least sixteen times if we roll sixty times in total and the bag does *not* contain a trick die. If we calculate that probability, known as the *p-value*, we see that it's quite small, only 1 in 30, but that's not a rare enough event for you to bet a month's salary that there's a trick die in the game. The more data you collect, the clearer the difference will be between the situations with and without a trick die. As the excess of fours becomes more and more obvious, at some point a coincidence will be so improbable that you *will* be willing to bet a month's salary.

But to bet your reputation as a scientist, you need a truly exceptional situation. We've agreed that we'll restrain ourselves until the chance of mere coincidence is less than one in a few million before we say there really is a loaded die in the bag. And that's exactly how we knew when to make a claim about whether or not the Higgs boson exists.

As with the trick die in the bag, we didn't know in advance whether the Higgs boson would cause an excess of certain measurements or, if so, where the excess would show up in the distribution of the invariant mass. What we did know was that if there was an excess of one of the two decay channels, then the other channel should show an excess too. That's because the peak in invariant mass should appear in both distributions around the mass of the Higgs boson. In fact, Peter Higgs's theory tells us what the relative strength of the signals should be, by predicting the relative probability of decay into two photons, compared to decay into four muons.

It was difficult to estimate the average number of collisions of each variety in the distribution if there was no Higgs boson—for two reasons. For one thing, this was the first time we were using the detector and the first time we were searching for collisions on this energy scale. In addition, the huge calibration program I mentioned earlier was still in progress. As we gradually learned more about how to identify the particles and their properties, we grew more certain of the average numbers of ordinary collisions (the background)—and so got better at estimating the significance of an observed deviation (a possible signal). Only after a lengthy series of checks could we finally look at the distribution of the invariant mass. To top it off, all this was deliberately encoded so that we wouldn't, consciously or unconsciously, work toward a particular solution. This meant, in part, that we didn't receive updates on the two distributions every day. But each update we did receive brought us a little closer to answering the question, "So is there a Higgs boson hidden in the data or not?"

Between the time when we saw the first excess and that moment of euphoria when the probability of a coincidental background fluctuation was less than one in a few million, the magical five-sigma threshold, we went through a long period of great suspense. You can imagine our excitement, keeping our eyes on that peak while the data poured in. But at the same time, it's nerve-wracking to keep checking whether the excess is increasing or decreasing each time new data are added. You have to wait until you can be certain, but the whole time, you know there's a competing experiment in progress, and those other scientists are just as eager as you are to claim the discovery. The spotlights of the international media, which want only a yes/no answer and not a complicated story about p-values and systematic errors, didn't make things any easier for those poor physicists.

Particle physicists work toward a few special dates every year where they announce their results to their colleagues and the rest of the world. The main events are the Rencontres de Moriond in March and one of the big summer conferences (ICHEP or EPS). The final weeks before the big summer ICHEP conference in Melbourne in 2012 were incredibly hectic. We expected to have collected enough data by that date to be able to make a new step forward in our research on the Higgs boson. CERN sent out a press release announcing an open presentation on July 4, 2012, by the leaders of the two largest experiments a couple of days before the beginning of the big conference. That's a very unusual step, and the rumors were flying—both in the laboratories and in the media. At Nikhef in Amsterdam, we set everything up so that journalists and the entire staff could watch the livestream of the CERN presentations, which were to take place in a large auditorium.

At CERN we're always rushing from one meeting to the next, but around that time it was an utter madhouse. Within the organization, we'd known for a long time that there was an excess in the data, but the results of all the calibration and checks trickled in slowly from all the different groups and had to be linked together before we could finally produce that one crucial number, proving that the peaks in the two channels were incompatible with the absence of a Higgs boson. To capture and combine all the data in one model so that, after hundreds of checks, we could reduce it to that one figure, the degree of incompatibility, we used a framework whose main designer and developer was my office mate at Nikhef, Wouter Verkerke. I still can't imagine how we ever would have managed to combine all that data so quickly without his computer programs. The compatibility with the hypothesis that the Higgs boson didn't exist had to be small enough—only then could we say we'd discovered the Higgs boson. And in that case, what was the Higgs boson's mass? Did the distributions provide a consistent picture? It drives us crazy at CERN that the results of major experiments always seem to leak out right away, but surprisingly enough, in this case we were able to keep the final numbers secret.

Although it's very hard for most people to believe, the final number was determined only a few days before the presentation. The excess that we saw was too great to be regarded as a fluctuation, and the rest of the picture was consistent too. So we had apparently produced a heavy particle whose properties were consistent with the particle predicted by Peter Higgs in 1964. We later learned that our competitors in the CMS experiment had also observed an excess, and in the same place. The particle, if it was the Higgs boson, had a mass of approximately 125 GeV (making it a little more than 130 times as heavy as a proton), and it did decay into two photons and four muons.

Physicists are very cautious about how they word their conclusions, and in the presentations and later publications, they described the measurement this way: "In the search for the Higgs boson, we see an excess consistent with the outcome that we expect if the Higgs boson exists, but there is still a very small probability that this is a chance fluctuation caused by other processes in the Standard Model." Ultimately, what you need is an authority figure to cut through the hesitation and turn a "maybe" into a "yes" or "no." In our case it was Rolf Heuer, then director of CERN, who said after the presentation by Fabiola Gianotti, the head of the ATLAS experiment (and the director of CERN as I write), "Ladies and gentlemen, I think we have it." The audience included François Englert and Peter Higgs, who had first thought of the mechanism and the particle itself. They had been invited to the seminar "because something interesting might be presented." It certainly was, and Peter Higgs showed his emotion when he was interviewed afterward, calling the discovery amazing and saying he was surprised it had happened in his lifetime.

A quest that took more than fifty years, completed with the discovery of the final piece of the puzzle by two large collaborative groups of scientists from around the world. Fantastic! It was a moment of euphoria, and also the end of the shared dream of finding that particle at the heart of the way we look at mass. How wonderful to have been a part of this historic breakthrough. As we suspected, empty space isn't really empty, but filled with the Higgs field that gives particles their mass. Not only the particles that carry the weak nuclear force (and whose sphere of influence is therefore restricted to very short distances), but also the very building blocks of matter. Mass is the property of particles that makes them cluster together and form structures in the universe, such as galaxies, the sun, and the earth. Now we understand what causes that. Some people have described this new reality as "like a fish discovering it lives in the water."

The New Reality

On July 5, like many of my colleagues, I woke up on cloud nine, but the nagging feeling that scientists know so well soon began to trouble us again. Sure, we had discovered a new particle, but did it have the exact properties predicted by Peter Higgs? Or was there a more complex mechanism by which nature assigned mass to particles, as so many theoretical physicists suspected? Once we had landed the biggest prize out there in physics, it may seem strange that we went on fretting about problems. You might almost forget, after our long obsession with the hunt for the Higgs boson, that the long list of unanswered questions in physics included many that could not be solved by the Higgs boson's existence.

We'll now look at a few of those open questions. We'll stay tightly focused on the phenomenon of mass, but all the same, there are still so many unsolved problems . . . even if that new particle really does turn out to be the one and only Higgs boson. For example, the Higgs mechanism explains *how* particles get their mass, but *why* they have their particular masses is still a complete mystery. And why are those masses so different from one another? A neutrino is probably a million times lighter than an electron, which in turn is a few hundred thousand times lighter than the top quark. Strange. Why are there three families that differ only in mass? And how is it possible that the celebrated Standard Model is perfectly consistent at the quantum level but fails hopelessly when we calculate the quantum corrections for the mass of the Higgs boson itself? In short, we still have plenty of work to do.

6

Terra Incognita
A Voyage into the Unknown

In our firm belief that the world is larger than we now know, scientists have tirelessly searched for new ways to extend the boundaries of our knowledge and explore new terrain. When we talk about mapping unknown territory we tend to think of explorers like the Chinese admiral Zheng He, but it's important to remember that scientists like Marie Curie also discovered and mapped new worlds. All were driven by the same sense of adventure and irrepressible curiosity. Our quest has brought us a series of new insights, and our knowledge about the universe, the earth, and biology today is incomparably greater than what we knew a century ago. In a book about science, it's tempting to focus on success stories, but we know it's really the questions we can't answer that point the way. After all, it's *out there*, beyond our present limits, that the answers to all our questions lie hidden. And it's those nagging questions that show us the world is larger than we now imagine. That was true a century ago, it still is today, and it always will be. We can now take the time to stare out over the edge into the mist and dream of what we might find there. There's a fairly long list of unanswered fundamental questions that rub our noses in the hard truth: the Standard Model is not the ultimate foundation of the laws of nature.

The world is larger, but where do we look, and which way do we go? Not one famous theory ever just popped into being out of thin air. A theory is almost always the result of a long search involving lots of trial and error. In the course of that long journey, ideas and dreams are promoted to laws of nature, and although failed adventures are all part of the game, alternative models that were unsuccessful tend to be forgotten, along with their creators. To understand what fundamental physics is searching for in my field, and what new worlds we're dreaming of, it's important to have a clear overview of what we *don't* understand.

It begins, every time, with collecting and organizing facts. That's how we gradually gain an understanding of nature, as each new fact is slotted neatly into a current model. That's all fairly boring work. The exciting part is when a new fact doesn't quite fit. And our blood really starts pumping when one of the following situations arises:

- We see something that contradicts our present laws of nature;
- We see a pattern that is not predicted by our present theory; or
- We pause once again to reflect on our list of "deeper" questions: the why behind the fundamentals of our theory, or issues like the origin of time and space itself.

In each of these scenarios, it's clear that our current picture of the world has fallen short and we need to expand the scope of our present laws of nature. But it's not the case that our old picture of the world then collapses and has to be thrown out completely. Absolutely not. After all, until the instant that inexplicable phenomenon arrived on the scene, the old model described every single phenomenon we'd observed in nature.

Sometimes a slight modification or extension of the old theory is enough to account for the strange new phenomenon, but that's not always possible. Some anomalies stubbornly refuse to reveal their secrets. In that case—that is, if the new phenomenon is truly impossible to understand—we eventually try to put together a new theory that sheds light on the problem from a very different or deeper perspective. This often requires us to introduce new concepts, new principles, new forces, or new particles. To an outsider, it may look as if we're just letting our imaginations run wild, but constructing a new theory is not as easy as it seems. That's because any new theory must meet a whole set of strict conditions. Specifically, it must:

- Explain all known phenomena and laws of nature;
- Solve the new problem/account for the new phenomenon (always a must, since that's the whole point of coming up with the new theory); and
- Avoid making predictions we know to be false.

We now see the creativity of *theoretical* physicists on display as well. Like architects designing beautiful and functional buildings, scientists coming up with completely new concepts and working out all their consequences are engaged in a creative process. They make sketches of a world that no one has seen, and those sketches—despite all the uncertainties— serve as guides for experimental researchers. Of all the possible "maps of the new world" made by theoretical physicists when they observe a mysterious phenomenon, of course only one can be right. The others will be stamped "incorrect" and tossed into the recycle bin as soon as their testable predictions turn out to be inconsistent with the experimental facts. After all, the facts are the final referee in the competition between the different theories. Even established theories are subject to regular stress tests, under the widest imaginable variety of experimental conditions. And in every case, no matter how good, how elegant, or how sophisticated a model is, as soon as it fails to explain an experimental observation it falls from its pedestal, and we start searching for an alternative.

The same way our search for answers to questions about the atom led to an explosion of new knowledge, we believe that now, once again, we are on the brink of a revolution. It's precisely because we keep learning more about the Standard Model that we're keenly aware of the things we don't understand, and it's becoming more and more evident that there's probably a more fundamental theory, of which today's Standard Model is merely a rough-and-ready version. Next we'll look at a few of the main trouble spots now pointing the way for further research in particle physics, and consider a few of the new and sometimes outlandish ideas that have been proposed as solutions to today's problems. Some ideas will sound deeply strange and radical, and although only one of them (at most!) will turn out to be true, that's no reason to take them less seriously. After all, the idea that does prove correct will turn our world upside down.

Some Phenomena We Don't Yet Understand

Here's a list of some big questions that keep particle physicists awake at night and point the way for research in our field. Our most wanted list:

1. **What is dark matter?** Although we pretend to know exactly what all the matter in the universe is made of (stars, planets, nebulae, neutron stars, and so on), we also know that we can identify only 16 percent of all the mass in the universe. Or to put it another way: for every known kilogram, there are four to five kilos of things, or stuff, or whatever, that we can't see. On top of that, we know this mysterious stuff does not consist of particles we have here on earth. So what is it? Or are we on the wrong track, and is there no dark matter at all?

2. **Why is gravity so weak and strange?** Of the four forces of nature that we've discovered, three are about equally strong at the level of the nucleus. They have the same mathematical structure and are all captured by a single quantum theory: the Standard Model. The fourth force, gravity, is still the odd one out. We just can't seem to find a way of describing its behavior at the quantum level. At the same time, it's very weak in comparison with the other three. That can hardly be a coincidence, but how do we explain it? Do we really understand how gravity works?

3. **Where did all the antimatter go?** In the Standard Model, matter and antimatter are in perfect balance. Whenever we produce a new particle at CERN, an antiparticle is always created along with it. So it's awfully strange that there's no evidence at all for the existence of antimatter anywhere else in the universe. Was there a slight excess of matter relative to antimatter in the

beginning, do matter and antimatter not behave in quite the same way after all, or are we simply "overlooking" half the stuff in the universe?

4. **What are those mysterious patterns in the Standard Model?** We also see weird patterns in the Standard Model that we can't even begin to explain:

   - Why are there three families of particles?
   - Why are there equal numbers of quarks and leptons?
   - Why do particles have the masses they do?
   - Why are the masses of neutrinos so incredibly small?
   - Do the three forces merge into a more fundamental, primal, force at high energies?

   . . . and many other, more technical questions.

5. **What about dark energy and the origins of space-time?** Meanwhile, let's not lose sight of the big picture. We've recently learned that the universe is not only getting bigger and bigger, but also getting bigger faster and faster. And that's odd. A cubic meter of space contains energy, so when the universe expands, that costs energy. What fuels this expansion? How do you create new space, anyway? And how did it all begin?

There are many other questions, but just looking at this short list, you can see that our voyage of discovery is not nearly over yet. And you can appreciate why scientists work day and night to come up with new experiments that may give us the answers. Of course, we haven't been sitting on our hands since we discovered the Higgs. In our search for answers, we've come up with various theories and experiments. I'll focus on three, which illustrate the three main areas of research: new particles (dark matter), new forces (the primal force), and new phenomena (extra spatial dimensions).

New Particles: The Search for Dark Matter

It's been clear for some time now that there's more matter floating around in the universe than we can see. The question of what this mysterious mass, dark matter, could be is one of the greatest open questions in physics today. One of the clues comes from the rotation of stars in galaxies and is easy enough to understand if we think back to the example of the earth revolving around the sun. We saw that if we knew the masses of the sun and the earth, we could calculate exactly how fast the earth has to move to remain in a stable orbit. Of course, we can see our sun with the naked eye, but imagine if that weren't the case. Even then, we could figure out indirectly from the movement of our planet not merely that there's something heavy at the center of the circle in which we move, but also *how* heavy that object must be. Since we know our own velocity and the size of our circular orbit, the laws of mechanics, which we've known for hundreds of years, tell us the rest. No room for doubt. Or so we thought. But to everyone's surprise, when we looked at the velocities of stars in galaxies, something was wrong. The stars' velocity didn't fit the distribution that we would expect. The only way for us to explain the distribution of velocities was by assuming that besides the stars we could see, there was also a cloud of matter particles present in each galaxy. Since we couldn't see those particles with our telescopes, we called them dark matter.

In recent years, indirect evidence has piled up for the existence of dark matter. Computer simulations show that if this mysterious extra mass were not there, ordinary matter would never have clustered together so quickly (that is, in the lifetime of the known universe) into the galaxies and larger structures we now observe. And there's other evidence for the existence of this dark matter. Even so, it's weird—how could we have missed such massive particles, when we thought stars, planets, interstellar gas, and photons were enough to explain it all? If we put together all the experimental results in cosmology and astrophysics, it's clear not only that the amount of matter we *can't* see is five times the amount of all known matter, but also—maybe more importantly—that it's not made up of the particles we find here on earth. Although the name *dark* matter seems to suggest that we don't have much information about its properties, all the clues we've found have actually taught us quite a bit. For one thing, we know it doesn't interact with ordinary light; otherwise, we'd be able to see it.

To make it clear how important this point is, I'll say it once again: the universe seems to contain five times as much matter we *can't* see as matter we *can* see, and all that unknown matter is *not* made of stuff we have here on earth. So what *is* that unknown matter? And if there's no place in the Standard Model for whatever particle it's made of, how can we expand the model to make room for it?

It seems so easy. If you're a particle physicist and you see new matter that doesn't seem to be made of the particles in the Standard Model, then you just think up a new particle. "That's what you guys do, right?" If only! That new particle would need to not only have mass, be electrically neutral, and hardly interact at all with ordinary matter (since those are the properties we've indirectly determined), but also fit into the structure of the Standard Model. Yet when you have a completed puzzle, there's no place for an extra piece. Likewise, we can't just force an extra particle into our current model. There *are* ways of solving the problem—tons of them, in fact. But they take us beyond the limits of that puzzle.

One promising theory that provides an elegant solution to the mystery of dark matter is known as *supersymmetry*. The concept was developed in the late 1970s, in an attempt to draw a connection between the building blocks of matter (fermions) and the force particles (bosons). Although the theoretical background would take us into too much depth for this book, one implication is that the number of particles in the Standard Model is doubled, because each particle is said to have a *superpartner*. Twice the number of particles—that sounds pretty drastic, especially considering that there's not a shred of experimental evidence for these partner particles. The only reason that physicists give this idea the benefit of the doubt is that the last time such a doubling was proposed, it miraculously gave us the right answer. That was back in 1928, when Dirac proposed the existence of antiparticles purely because of the mathematical structure of his theory.

Dirac, too, doubled the number of particles in one stroke, and four years after his prediction, the antielectron was actually discovered. What do you know! Score one to zero for the theorists, as the opening goal in a game that would go on for a hundred years. We now know that Dirac was right: each particle really does have a corresponding antiparticle with exactly the same mass. So should we consider the possibility of another doubling? Is there a mirror world, and if so, why haven't we discovered it yet?

In the simplest version of supersymmetry, the superpartners had the same mass as their counterparts made of

ordinary matter, and since we knew the properties of those ordinary particles, it was clear that we should have seen the superpartners in our experiments a long time ago if they really existed. But we soon found out that a slight modification to the theory permitted the new particles to be heavier than their partners in the everyday world after all. Suddenly the idea of supersymmetry was alive and well again, because in the new version, the particles from the mirror world really do exist—they're just too heavy to be produced at the collision energies we've used in our experiments so far. But—and this is what really gets particle physicists worked up—there's a good chance they're light enough that we can prove their existence in the collisions in CERN's Large Hadron Collider.

Supersymmetry is a sketch of a new world, and now that we physicists have been living with the superpartners for a while, they've begun to seem almost familiar. We've even made up names for them. For instance, the selectron and the squarks are the partners of the electron and the quarks, the wino is the partner of the W boson, and so forth. Out of all these strange names, there's only one you need to remember: the *neutralino*, the most famous new particle in this model. This is the particle that physicists believe drifts around the universe in huge quantities, forming dark matter.

Like almost all elementary particles, these supersymmetric particles don't live forever. If you made them in a particle accelerator, they would, like the heavy particles in the Standard Model, decay step by step into other, lighter, particles until only the lightest stable particles remain. But here's the exciting part: the theory allows for the possibility that the lightest of the supersymmetric particles (the smallest of the new giants) is stable. That implies that once this particle has been created, it cannot decay any further on its own, so it'll go on whizzing through space until the end of time. Hey, that's just the kind of thing we've been looking for!

If we "simply" assume that the number of particles is doubled, and that each new particle has its own large mass, the number of imaginable models is truly enormous. In the most general case, there are many dozens of parameters that can vary freely (the masses of each new particle, for one) and each set of choices leads to a unique mirror world and to different predictions about what we will observe in our experiments. It's up to the creators of new models to investigate which scenarios are consistent with established experimental facts and which are not. For example, they shouldn't predict particles that would already have shown up in our experiments, and their specific model's prediction of how many dark matter particles are now whizzing around the universe should fit perfectly with our observations. Over the years, many models have been excluded this way, but unfortunately many, many more are still possible. We have to hold out for a measurement that gives us something to go on and shows that something like supersymmetry exists. The hope of finding these supersymmetric particles, especially the particle responsible for dark matter, is now one of scientists' most powerful motivations, at CERN and other laboratories around the globe. This is too important for us *not* to try out a range of hunting methods. So that's what we're doing.

There are many reasons to hope that supersymmetry is real, but most importantly, it would give us a candidate particle for explaining dark matter. That explains why, despite the vast array of possibilities, we're looking for evidence from all possible directions. Next I'll describe a few experiments under way to prove the existence of dark matter particles. Each of these strategies is intended to provide an answer within the next few years. It promises to be quite a race.

MAKING DARK MATTER IN A PARTICLE ACCELERATOR

If you've learned anything in the course of this book, it's that you can produce new particles in an accelerator. And it's true: just as you can make Higgs bosons by smashing protons together with enough energy, it should also be possible for similar collisions to produce supersymmetric particles. That is, if they exist.

If we do manage to produce these heavy particles, they're certain to fall apart again right away, and in the debris we would expect to find not only the familiar particles, but also the particles of dark matter. All of this would take place in the lab under controlled conditions, and only at energy levels high enough to produce those heavy particles, of course. And even then, supposing you *do* produce them, it's far from trivial to spot them among the billions and billions of "ordinary" collisions that take place every second. The only way to identify them is the same way we did the Higgs boson: by looking for the unique fingerprint that clearly sets them apart from all other particles.

Finding a pink sheep in a big herd is not nearly as difficult as finding the one sheep with a black spot on its belly. How big is the spot supposed to be, anyway, and did I take a close enough look? Each of the many theoretical models enables you to predict what you'll see in the experiment. Those predictions range from a pink sheep ten times the usual size to sheep that *may* have a tiny spot almost the same color as its skin. It's crucial to come up with the right strategy for recognizing the possible signal and filtering it out of the large data set—just as we did in our search for the Higgs boson.

One of the universal elements in all these new theories is that a collision producing supersymmetric particles always creates not only ordinary particles but also particles of dark matter, which end up in our detector. The frustrating part is that we know we can never see these particles in our detector, because they hardly interact with the other particles of which the detector is made. Despite the fact that there's measuring equipment all around, they'll fly straight through it. So how do you measure something you can't measure?

That all sounds horribly difficult, but we have a strategy: we use what are known as the laws of conservation. In a prison, the wardens can easily tell whether anyone has escaped by comparing the number of prisoners at morning roll call to the number who were locked in their cells at bedtime the night before. If there are fewer prisoners in the morning than went into the cells at night, then one of them must have escaped, even if you never noticed anyone climbing the fence. Collisions in a particle accelerator work exactly the same way. In a full-on, high-speed collision between two particles, there is no energy perpendicular to the direction of motion. So when the fragments fly off in all directions, equal amounts of energy go to the left and to the right. And it's easy for us to measure that energy. A perfect balance. Unless, of course, a particle of dark matter escapes, because then you can't see it.

You might compare it to a tomato dropped straight to the ground from a couple of meters up. When it hits the ground, it bursts apart and makes a large circular splotch. In this case, too, equal amounts of tomato pulp splatter to the left and to the right. In a particle collision, part of the tomato pulp is invisible (the particle of dark matter). Like the falling tomato, the collision should make a symmetrical pattern, so if we observe an asymmetry, it suggests that a particle escaped detection. So we can prove indirectly, without seeing it, that something escaped—in other words, that our collision created an invisible particle.

Unfortunately, it's not quite as easy as I make it sound. There are quite a few complications. You first have to know for certain that your equipment can measure all the particles accurately, and we also know that some ordinary particles—namely, neutrinos—can escape unnoticed. So it's vital to make absolutely sure that your detector works and the observed asymmetry is real. Our efforts to analyze and interpret images of collisions with confidence have kept hundreds of scientists hard at work for more than ten years. For the reasons we've just discussed, many of my colleagues at the Large Hadron Collider are in search of collisions where the image clearly shows that something is missing. And if there *is* something missing, then they have to look for evidence that, in this case, the initial collision really did produce supersymmetric particles. In terms of our prison roll call, the question is: which prisoner escaped?

Although a large group of scientists has been at work night and day at the LHC in recent years, searching for evidence of these new particles, we still haven't found any sign that any particle was created that escaped our attention. Fortunately, this kind of negative result can have a partly positive impact, because the fact that we *haven't* observed something can lead to new insights. You see, if supersymmetric particles exist, then this negative result in our search for those particles means *either* that they're too heavy to produce at the energies available in our collisions at the LHC, *or* that there's a conspiracy: nature set the parameters of the model so that the fingerprint is not unique enough for us to tell the crucial collisions apart from ordinary ones. Both possibilities would be very unfortunate, but there's hope. The thing is, we also know that so many particle collisions will be produced in the years ahead that even very subtle clues will become visible in the data. We're really excited about it. After all, there are not many other theories that would solve as many problems as supersymmetry, and a lot of scientists would sleep better if we proved it correct. But however much we'd like that, only our experiments will tell us if it's more than just an idea, or if we need to start searching in a different direction.

COLLIDING PARTICLES OF DARK MATTER
IN OUTER SPACE OR IN THE SUN

The main reason that the existence of supersymmetry is believable at all is that, according to the theory, the lightest of the new particles is stable. That would explain why there are still so many particles of dark matter drifting through the universe now, billions of years after the Big Bang. Even though they can't decay on their own, there *is* a little loophole that would allow particles of dark matter to disappear gradually: the same theory predicts that when two particles of dark matter collide, they produce two ordinary particles. Those could be two photons, two neutrinos, or an electron and an antielectron. If we look at parts of the universe with a lot of dark matter in one place, we should be able to see a signal (in the form of light) that is characteristic of the annihilation process when particles of dark matter meet.

Some people think there may be a lot of dark matter in the center of the sun. It's only logical, they say: since dark matter particles are heavy and pass straight through ordinary matter, they should gradually sink toward the center of the sun. So the scientists who believe this actually look straight into the sun, for professional reasons, to see whether it's sending out a slightly stronger neutrino signal. They've been fairly successful at measuring the signal, but so far it doesn't seem any stronger than expected. Too bad!

A different tactic is to use satellite experiments to look at differences between signals from parts of the skies where there are thought to be especially large or small amounts of dark matter. The Fermi satellite experiment has claimed to have found a difference in the numbers of photons coming from different parts of the sky, depending on how rich or poor in dark matter they are. The PAMELA satellite experiment and the AMS experiment also seem to have found indirect evidence that more antimatter (in the form of antielectrons) is produced than you would expect if no dark matter particles ever collide in the universe. The AMS experiment is quite special. It is a miniature version of the particle physics detectors we discussed earlier in the book, mounted on the International Space Station. The goal of the international experiment, led by Samuel Ting—a Chinese American physicist and Nobel Prize winner—is to measure the amount of antimatter in the billions and billions of cosmic rays that pass through the detector. A few of the tantalizing hints from these measurements could also be interpreted as coming from other phenomena, but even so, these are exciting times. It feels like we're hot on the heels of supersymmetry.

DETECTING COLLIDING PARTICLES OF DARK MATTER DEEP UNDERGROUND

Aside from producing particles of dark matter in the lab, there are other ways we can detect them. For one thing, if there really is dark matter floating around in space, then some of it is probably close to the earth. And since the earth and the solar system are flying through space, we must constantly be moving through a cloud of those dark matter particles. That doesn't really bother us too much, since dark matter can pass through solid rock the same way we fall through the air. Even so, there's a chance, however slight, that a particle will "bounce" off an atom's nucleus if it gets close enough. It's unlikely in the extreme, but never mind. As long as there's some chance, however small, experimental physicists will find a way to make something of it.

An atom hit by dark matter picks up an electrical charge, and with the help of an electrical field, the charged debris can be sent to a special detector, where it generates a slight current. The chance of a collision depends on the number of particles in the atom's nucleus: the larger the nucleus, the greater the probability. And if you want the remnants of the atom to be able to move through the surrounding substance, you need to use a gas or a liquid. Colleagues of mine went in search of a very heavy liquid and eventually settled on fluid xenon. Two thousand kilograms of xenon, to be exact, and since it's three times as heavy as water, that comes to about six or seven hundred liters.

In principle, all you have to do once you've built your detector is wait and see how many collision signals you detect as the earth passes through the sea of dark matter. If you calculate how many collisions you would expect, you soon find it's no more than a handful at most. Per year. That could be enough (since you need to see only one Smurf to prove that they exist), but only if you can rule out the possibility of a false signal: a signal caused by an ordinary particle that you wrongly interpret as a collision with a dark matter particle.

Physicists are by nature very skeptical, so before they claim to have seen "new physics" in their own experiments, they want to be sure they haven't made a single mistake. The two strongest effects that could produce a false signal are (1) the natural radioactivity in the rock and metal of the huge "thermos" in which the liquid xenon is stored, and (2) cosmic radiation. To neutralize the first problem as much as possible, the materials of the detector have to meet ultra-high quality standards. The wrong kind of glue, containing a trace amount of radioactive material, could ruin the whole experiment. But however difficult it may be, at least you can do something about that problem.

It's harder to deal with cosmic radiation, rays from outer space that hit the earth and produce the muons we talked about earlier. In this experiment too, those muons will show up in your detector. To shield it off, you'd need hundreds of meters of stone. But that's impossible. Isn't it? Well, not if you run your experiment deep underground. And there are plenty of places in the world where that's possible, like abandoned mines, or mountains through which we've drilled tunnels. You'd be surprised how many mines have been taken over by physicists, and they're doing experiments under mountains too. Imagine if you could carve a vast chamber into the solid rock beneath a mountain. It would be the ideal place for this kind of experiment—so that's just what scientists have done.

It sounds a little like *The Lord of the Rings* or a James Bond film: spaces the size of a cathedral, hollowed out of solid rock. But they really exist. Most people don't notice as they drive through the Frejus tunnel in France, or the tunnel through the Gran Sasso near Rome, that there's an extra lane for a little while in the middle, with a boom barrier and a gate that opens onto a road leading deeper into the mountain. This road leads to an underground experimental laboratory, a place where your experiments are shielded from cosmic rays, but where you still have access to computers, electricity, gases, climate control, and so on. You have to go a day without sunlight once in a while, but that's the kind of sacrifice a scientist is more than willing to make.

The Netherlands is involved in one of the experiments in progress in the Gran Sasso lab: the Xenon1T experiment. If we really are flying through a sea of dark matter particles, and if those particles have more or less the right properties, then this kind of experiment has a good chance of spotting them. That would be fantastic. Not only because the experiment is running at the same time that the LHC accelerator collisions are being analyzed, making the two experiments complementary, but also because I know how hard my colleagues have worked and how much their success would mean to them.

PRECISION MEASUREMENTS AND
QUANTUM CORRECTIONS

The final way of searching for the impact of the new supersymmetric particles is to look for their subtle effects on all sorts of calculations. The calculations you have to perform to make predictions in particle physics are never a matter of simple arithmetic. In practice, a prediction in a quantum mechanical theory is the sum total of *every* possible thing that could happen. And in quantum mechanics, a lot of strange things can happen. In fact, whatever *can* happen *will*—though sometimes only with a very low probability.

For example, one of the strange predictions of quantum theory is that for just an instant, you can (to put it much too loosely) "borrow energy from the vacuum." Then you can use this energy, very briefly, to make a set of those new, heavy, supersymmetric particles. So even though there's not actually enough energy available in the process to make them "for real," the easy loan terms offered by the vacuum (you can borrow as much as you want, but the more you borrow, the sooner you have to return it) allow you to make two heavy particles for a split second. The interesting thing is that these *virtual particles*, however briefly they may be part of a physical process, still have an effect on the outcome of the calculation. It's a small effect, but if you measure precisely enough, you can determine the size of these quantum corrections and may be able to prove indirectly that the new particles, for one fleeting moment, really were created. And therefore existed. This method has been used a number of times to prove that a Standard Model particle existed, before it was produced "for real" in an experiment.

The same possibility exists for the supersymmetric particles. Some processes at the LHC are very sensitive to the presence of new particles of this kind. One example is the very rare process in which a B_s particle (consisting of a bottom quark and a strange quark that go through life together) is created and then decays into two muons. The Standard Model predicts that this will happen to three B_s particles in a billion, but if a relatively light supersymmetric particle existed, then it could conceivably happen many times more often. The first measurement, however, in which colleagues of mine from Nikhef in Amsterdam played a central role, yielded a result exactly identical to the Standard Model prediction. A fantastic measurement, sure—a triumph for the Standard Model. But at the same time, a bitter disappointment for the supersymmetry hunters. There are still huge uncertainties, but that was not a good day for supersymmetry. Fortunately, this is just one of the possible measurements that could provide evidence for it. In the LHC experiment, a feverish search is under way for any anomalies that may suggest new effects impossible in the Standard Model. Achieving the level of precision required is an incredibly difficult task, but the experiment is doing an amazing job.

The origin of dark matter is one of the greatest mysteries we face. It's a central component of the way we describe the evolution of the universe: phenomena like the formation of stars, solar systems, and other large structures. Since it's clearly not part of the Standard Model, it is one of the clearest signs we have that our current model falls short and that the right theory has not yet revealed itself. So you can see why dark matter is being hunted from all sides. We're trying to make our own supersymmetry particles in accelerators and observe their effects indirectly through precision measurements. At the same time, we've burrowed underneath a mountain to look for collisions of dark matter particles with our detector, and we're even using satellites to search for collisions of these particles with each other. A variety of experiments, with a variety of approaches, but they also have a common objective and a common time frame: each will complete its measurements within five years and come to a conclusion about dark matter. We have incredibly exciting times ahead, and if one of those experiments finds an anomaly or a signal, you'll read about it on the front page of every newspaper.

New Forces: In Search of Unification

When we discussed the forces in the Standard Model, we saw that those forces can each be described in terms of an underlying mathematical symmetry: $U(1)_Y$, $SU(2)_L$, and $SU(3)_C$, to be exact. These symmetries "explain" not only the existence of the force carriers (and the number of varieties), but also the way ordinary particles exchange those force carriers. Although the forces differ greatly from each other in structure and in the resulting phenomena, we also see a picture emerging in which not only is each force based on a mathematical symmetry, but there is also one distance, much smaller than anything we can see with our current particle accelerators, at which the strength of all those forces is approximately equal. This combination of factors seems to suggest the possibility that the forces known to us today could have developed out of a single primal force. This unification of forces is also known as the Grand Unified Theory (GUT). But how does it work exactly, what makes it so interesting, and how can we find out if it's true, given that we aren't yet able to study nature on that much smaller scale?

If we hold onto the direct link between symmetries and force carriers as our guiding concept, the existence of a primal force of this kind could be the result of a primal mathematical symmetry—a complex symmetry, in other words, of which those in the Standard Model are just simple parts, much as the rotational symmetry of a circle is a simple form of the higher rotational symmetry of a sphere. If you dive into the mathematics of symmetry groups, you soon realize that there are an astonishing number of ways to come up with a broader symmetry into which the symmetries of the Standard Model would fit. Arguments can often be made for or against a particular new symmetry, and each possibility has its own implications for the new phenomena that will show up in experiments.

One of the most popular examples of a higher symmetry that leads to a primal force unifying all known forces is what we call the SO(10) group. A unification of this kind is much more than a simplification that takes you from three separate symmetries (and therefore forces) to a single unified one. One interesting implication, for example, is that *all* the particles in each family truly belong together. Just as in the Standard Model the symmetry of the weak nuclear force leads us to think in terms of electron/neutrino and up-quark/down-quark pairs, in a model like this one *all* quarks and leptons can be brought together in the same structure. Then they truly belong together, forming a tight-knit family with ten members.

This final aspect would resolve a large number of questions. For instance, in a structure like that it would make perfect sense that there are equal numbers of leptons and quarks, two worlds that are completely separate in the Standard Model. But as I mentioned, there are also disadvantages. One implication of a tighter connection between quarks and leptons is that this new force would be able to transform them into each other. That's analogous to the weak nuclear force phenomenon in which the W boson can transform elements of an isospin doublet into each other: electrons and neutrinos, or up quarks and down quarks. The new possibility revealed by the new symmetry is that quarks and leptons can be transformed into each other as well.

And we really don't like that idea. That's because it would mean that a quark inside a proton can transform into a lighter lepton, such as an electron. Is that a problem? Yes! You see, it would imply that protons don't live forever. But all the evidence seems to show that they definitely do. Earlier, we talked about the fact that experiments have shown protons to have a lifespan of at least 10^{34} years—much, much longer than the age of our universe. I should add that there's always a theoretical escape clause, a way of magically making everything work out by very carefully tuning specific elements of the model. If you do that, things soon get complicated and less "elegant," but it *is* possible.

Before we go on to discuss how we're trying to prove the existence of the unified force experimentally, let's pause and fully appreciate these amazing measurements of the lower limit to the almost everlasting life of the proton. These measurements point the way within the development of numerous theories—a fine example of the interplay between theory and experiment.

As I mentioned, experiments have showed that the proton's lifespan must be many times greater than the age of the universe. But how can you ever prove that? You can't watch a single proton for billions and billions of years in your lab to see if it falls apart, can you? No, you certainly can't, but you could spend a few years watching billions and billions of protons to see if any of them fall apart. A convenient trick! One experiment, using the Super-Kamiokande detector, located a thousand meters underground in the Mozumi Mine in Hida's Kamioka area in Japan, looks at proton decay in a huge tank of very pure water (which is intended mainly for a different measurement). This has taught us that if a proton ever decays, it must take longer than 1×10^{34} years. That's about a million times a billion times a billion times the lifespan of the universe. In other words, a good long while. The ultimate neutrino detector, DUNE, now in construction in the United States, will also make the search for proton decay an important part of its impressive research program.

Back to the idea of the primal force. Although we aren't yet able to investigate the small scales on which this new primal force reveals itself, we fortunately do have a way of saying something about it, namely by proving the existence of the Z' particle. If the primal symmetry is composed of lesser symmetries, then our mathematics tells us that besides the three fragments of symmetry we now observe—$U(1)_Y$, $SU(2)_L$, and $SU(3)_C$—there should be another (small) piece. This piece—a single $U(1)_X$ symmetry—would, as with the other forces, have to imply the existence of an extra force carrier. In that case, it should be possible to produce and detect that massive force carrier in the proton-proton collisions at CERN, just as we produced the Z boson from the Standard Model. That's only if it exists, of course, and only if it decays into particles that we can see in our detectors.

The search for this Z' particle, which is high on our most wanted list of fugitive particles, is an area of research now getting a lot of attention in experiments at CERN. Not only because of our interest in the new force of nature, but also because it would be relatively easy to identify in collisions. If the LHC can squeeze enough energy into a collision to produce the Z' particle, then it may decay into particles that are easy to recognize with our detector—for instance, two muons or two electrons. More than once in this book, we've run into a method by which we could then prove the Z' particle's existence: namely, by looking for a slight peak in the distribution of the invariant mass of the two leptons. In other words, another search for the peak. But unfortunately, the collisions at CERN have not yet turned up any evidence of the existence of this new Z' particle.

As you can imagine, the absence of a peak doesn't bring a smile to anyone's face, but we're almost as hopeful that we'll find it in the near future as we were before the start of the LHC. That may seem strange, as if we won't admit defeat and are still clinging to an idea that turned out to be wrong—but the absence of a signal can't be linked one-to-one to the truth or falsity of the unification of forces or the existence or non-existence of a new force. Why not? Well, how often the Z' particle is produced, and whether it produces a visible signal in the detector, depends on that particle's properties—which are unknown. Maybe it does exist, and we just didn't have enough energy in the collision to produce it. Or maybe we had enough energy but it couldn't be produced at the LHC because it never couples to quarks or gluons. After all, we don't know what "charge" the new Z' boson couples to. And if the particle strongly couples to quarks, that could be a problem too, because then it won't decay into leptons and show up as a peak in the distributions. All these properties of the Z' particle depend heavily on what new, higher symmetry you come up with. And no one knows what symmetry nature has chosen.

The unification of forces (and the related unification of leptons and quarks in some models) is still a very attractive idea. I'm a big fan of it myself, although to be honest, it's hard to stay as enthusiastic after failing to find a definite signal.

New Phenomena: Extra Dimensions of Space

In physics, gravity is the odd force out. Albert Einstein's general theory of relativity gave us a way of describing the dynamics of space and time. The theory can not only describe falling apples, the earth's orbit around the sun, and clustering galaxies, but can even predict the existence of black holes and gravitational waves. Yet even though the theory of relativity is a magnificent creation that has been used successfully for more than a century, we physicists still have an uneasy feeling about it. Besides not really understanding why space should curve and cause objects to attract each other, we're also puzzled by a problem on a smaller scale. At short distances, three of the four forces of nature can be described by a single quantum theory, the Standard Model, but we just can't seem to come up with a similar quantum theory for gravity. Gravity can't be tamed.

You also notice something weird when you compare the strength of gravity on that small scale (the atomic level or below) with that of the other forces. The three quantum forces —electromagnetism, the weak nuclear force, and the strong nuclear force—are all about equally strong at that distance, but gravity is far, far weaker. How much weaker? Take a deep breath, here we go . . . gravity turns out to be a hundred times a million times a million times a million times a million times a million times a million times as weak as the other three forces. Of course, it's possible that's "just how it is," but we strongly suspect we're missing something. There must be some deeper explanation, and we must be overlooking something big, the final piece of the puzzle. But what?

I discussed earlier hidden properties that explain certain things as soon as you discover them. Two people who look the same at first glance (same age, same gender, same educational background, same number of children, same neighborhood) are likely to have similar ideas about a lot of things, but may react to the same situation in totally different ways. That could have any number of simple explanations. Maybe one was just fired and the other promoted, or one is blind and the other is deaf, or one is a devout Christian and the other an atheist. If you don't know about these kinds of individual factors, you won't be able to explain why they react so differently. We often say that this information adds an extra dimension to the situation we're looking at. I think we've all gotten ourselves into sticky situations when we didn't know all the facts about the person we were talking to. Whether you can talk your way out of those situations depends on your verbal skills, but in general, it seems fair to say that it's crucial to know all the parameters at play before you can truly understand a situation.

Physics is no different. The history of science offers many examples in which using a new perspective to reframe a familiar, complex problem suddenly made it very simple. Even logical. Describing the orbits of the planets and the sun in our solar system, while assuming the earth was the center, was very complicated. Then someone realized the sun might be the center. A new insight like this one often comes from a single person, and in hindsight, it seems so obvious that we wonder why no one ever saw it before. The same applies in other areas of physics: time after time, observations of inexplicable phenomena have added a new dimension to our understanding of nature. That extra dimension could be a new way that particles behave (think of quantum mechanics), a new phenomenon like the theory of relativity, a completely new force (like the strong and weak nuclear forces), or even a previously hidden property like electron spin.

In physics, there's also a more literal meaning of the term "dimension": a degree of freedom of movement in space. A line is a one-dimensional object because you can only move in one direction on it, whether forward or backward. A plane (such as a tabletop) is two-dimensional because you can also move to the left and to the right. Throw upward and downward motion into the mix, and we're in three-dimensional space. That's the kind of space we're used to, without any other degrees of freedom. Sure, there's also a fourth dimension, namely time— in which you can only move forward, to many people's frustration. But there aren't any other spatial dimensions. At least, not that we know of.

In 1998 Nima Arkani-Hamed, Savas Dimopoulos, and Gia Dvali published a paper entitled "The Hierarchy Problem and New Dimensions at a Millimeter." Another paper on this subject, just as well known, is "A Large Mass Hierarchy from a Small Extra Dimension" by Lisa Randall and Raman Sundrum, which dates from 1999. The paper by Arkani-Hamed and his colleagues was one of those rare, truly groundbreaking publications that introduces a simple, strange new concept offering a fresh perspective on one of the biggest questions in physics: in this case, "Why is gravity so weak, and why does it act so weird?" They added a new dimension in a very literal sense, proposing that there may really be an extra spatial dimension. At first that may seem fairly radical, or even nonsensical, because—putting aside the fact that we have to wonder now and then how there can be such a thing as space at all (yes, scientists are seriously trying to answer that question)— we can hardly imagine how there could be another dimension besides the three familiar ones. And it's not clear at all how an extra direction would explain the weakness of gravity compared to the other forces of nature.

Before I explain their idea in more detail, let me sketch an analogy that may shed some light on what makes the concept of an extra dimension so powerful. If at the end of the next page, you find yourself mumbling, "yeah, sure, I get that," then I'll have achieved my goal. Then you'll have grasped one of the most abstract new concepts in physics, and that's well worth bragging about at birthday parties or over dinner.

If you're having a drink with friends in the garden one summer day and an ant walks across the table, you know that it's stuck on that tabletop. Physicists and mathematicians say that the ant lives in two dimensions. After all, it can only move in two directions: left/right and forward/backward. It doesn't have any other choices. Though you wouldn't normally stop to think about it, it's an interesting and instructive exercise to imagine yourself in the ant's place. How does *it* see the world?

But before you start empathizing with the ant, let's add one more detail to our story: the table is slightly crooked, with the right side a little higher than the left. Now, there may not be much going on in an ant's head, but when it walks across that tabletop, it notices pretty fast that going toward the right side takes more energy than going toward the left side. Since the entire tabletop looks the same to the ant, no matter what direction it's facing, that difference between moving left and moving right seems strange and illogical to it. But it has to face the fact that the right side is harder to walk toward than the left

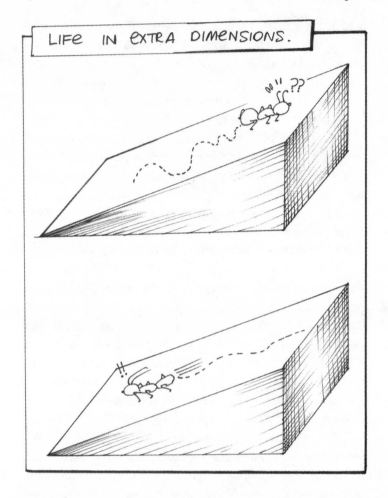

and learn to live with it. Period. A brilliant ant might be able to predict that if it spits out a grain of sand, it will always roll to the left and not to the right, but it could never figure out *why*.

We humans *do* know the reason, of course: the table is crooked. But even if you could talk to an ant, it would be hard to convince it that an extra dimension exists. Consider the ant's point of view: it can't enter that up/down dimension, it can't look at the table from the third dimension the way you can, and your explanation could just be a mathematical trick intended to pull the wool over its eyes. How can it figure out whether the third dimension is real or an illusion?

One way to convince the ant is to show it that the "trick" also solves other problems in its world. A smart ant will also be puzzled by some other things it's run into on its journey across the table. How is it possible that some objects change in size when he walks around them, and why do objects appear and disappear, seemingly out of nowhere?

This question asks a lot of your imagination as a reader, but you've come so far in this book now that I trust you'll take this last step with me. Let's think about what an ant sees, exactly, as it walks across the table, unable to look up. Squat down for a moment and look just over the edge of your desk, holding a sheet of paper out horizontally just above your eyes and the table like a visor, so that all you see is a thin strip. From this ant perspective, if you look at a round glass on the table, all you see is a small stripe, and if you walk around the glass (remember, an ant doesn't know it's a glass and sees only the stripe), the obstacle stays equally broad. No matter what side you look at it from. Logical for a human, and maybe for an ant as well. As for those strange objects he sees changing in size, if one of the glasses on the table is square, then it matters from which side the ant looks at the glass. If it faces a side straight on, the stripe is shorter than when it faces a corner. That's because it then sees the diagonal, and the stripe looks about 40 percent longer. A strange experience for our ant. So as it walks around the object, it sees a stripe that keeps getting longer and shorter for reasons it can't explain. Very odd. If you then make that object suddenly disappear and reappear, simply by picking the glass up and putting it back down, even the most skeptical ant will be convinced: there really is an extra dimension. At that moment, the ant will realize that the world is much bigger than it always believed and that people could observe that ant from close by without it noticing.

A mathematical problem is not often described in the form of a novel, but that's exactly what Edwin Abbott Abbott does in his book *Flatland*. That astonishing book, written more than one hundred years ago (and still available everywhere) describes the world of two-dimensional creatures such as triangles, squares, and circles, one of which is visited by a sphere—a creature from the third dimension, which introduces him to that new world. At the end of the book, the two-dimensional creature asks the sphere if there could also be a fourth dimension. The sphere calls the idea absurd, because he can't picture it. But that's exactly the question I'd like to ask you, my reader, and which I've been working toward for two pages now.

Try to imagine there's an extra dimension of space—the fourth dimension. Like the ant, we're stuck here in our three-dimensional world, but there's a whole world outside us we can't get into. Not easy, right? According to the authors of those late twentieth-century theoretical papers I mentioned earlier, if you're able to take that step, then you'll realize, in a flash of insight, why gravity is so weak. From that perspective, it's perfectly logical.

HOW THE EXTRA SPATIAL DIMENSION EXPLAINS GRAVITY'S WEAKNESS

Why would an extra dimension solve our gravity problem? Well, it would actually change our picture of the world in many ways. The same way an ant may never know that you exist even though you're leaning over it and observing its life, it's perfectly possible that there's a whole world very close by that we can't perceive. A world of unimagined possibilities. Even if it's not true, it's not only a goldmine of inspiration for philosophers and science fiction filmmakers, but also just screaming to be picked up by therapists who can give it an exotic, New Age twist. When I heard this concept of extra spatial dimensions for the first time, I thought it was utter madness, another typical mathematical trick being played by my theoretical colleagues. After all, we haven't observed any extra dimensions. But the scientists making this proposal have impressive track records, so my colleagues and I hit the books and then put our heads together, trying to fathom the idea, the mathematics, and the consequences.

Here's what it comes down to. If there were an extra spatial dimension, it would have to have the extra property that we, with our ordinary particles and forces, can't normally enter it. That would be why we don't see it in everyday life. Apparently, we're just like ants, stuck on our three-dimensional tabletop. But now comes the big idea: suppose gravity can see that extra dimension of space. That means that while the three other forces of nature move only in three dimensions, gravity has to radiate into the full four dimensions. So here on our paper-thin three-dimensional tabletop, we feel only a tiny fraction of the full gravity acting on the full four dimensions. Bingo, problem solved! Well, okay, but is it true or not? In the end it's up to us three-dimensional experimental physicists to prove that it is. Or isn't. They've been known to be wrong, those theoretical physicists. And they have such active imaginations.

THE EXPERIMENTAL SEARCH FOR EXTRA DIMENSIONS

This theory of extra dimensions can have dozens of variations: infinitely large dimensions or very small ones, theories in which our forces can partly leak into the extra dimensions, and so on. And like the ant, we want to see proof that they really exist—or that we're just chasing a mirage. So the question is, if they exist, how could you see them? After all, all our forces (including light and our particles from the Standard Model) can't enter that extra dimension. But there is a solution. Every version of this theory implies that at very small distances gravity will reveal its full strength. The big particle accelerator at CERN is a powerful enough machine to bring us a big step closer, allowing us to observe smaller scales than ever before.

What are the clues to extra dimensions that we should be looking for in all those collisions? There are two: gravitons and mini black holes.

If this theory of extra spatial dimensions is correct, and gravity really becomes just as strong as the other three dimensions at very small scales, we should be able to observe the quantum mechanical effects of gravity—and more specifically, gravity's force carriers: *gravitons*. If we can bring together enough energy at CERN in a proton collision, then the same method we use for photons and for W and Z particles should allow us to produce a graviton. That particle could then escape into the extra dimension of space and make energy suddenly disappear. It could also reappear again, just as suddenly—the same way that, to an ant, a glass disappears when picked up from the table and reappears when put down somewhere else. Energy, seeming to disappear into nothingness.

One of the strangest implications of the theory of extra dimensions—in which gravity can become incredibly powerful at short distances—is the possibility of creating *mini black holes*. You see, gravity can give rise to strange phenomena, because of a property that other forces don't have. Much like electromagnetism, gravity becomes stronger as particles come closer together. But then something strange happens. At a certain threshold distance, the Schwarzschild radius (you might call it the point of no return) named after the German physicist and astronomer Karl Schwarzschild, gravity becomes so strong that it creates a black hole: an object from which not even light can escape.

That sounds dangerous, because all the black holes in books and TV series about the cosmos are extremely heavy objects that suck in everything around them, sort of like cosmic vacuum cleaners. If this theory is right, would it mean that we can exceed the limit in particle collisions at CERN and make mini black holes? "Better safe than sorry: don't even try it," you might think. And that seems fair enough: after all, wouldn't the black hole swallow up the earth? CERN has taken public concerns like these very seriously and studied all possible doomsday scenarios for the Large Hadron Collider in great detail. That includes the phenomenon of mini black holes.

Here's the executive summary of that research: there's no danger at all. Even though the LHC smashes together protons at the highest energies ever generated by humans, there are proton-proton collisions in our universe with much higher energies. In fact, they take place daily in our own planet's upper atmosphere. We talked earlier about cosmic rays that hit the earth. Those are protons crashing into the nuclei of atoms high in our atmosphere. The highest-energy cosmic rays carry collision energies many times greater than those at the LHC, and as we can see, the earth—after billions of years of bombardment with cosmic rays—is still around.

There's also a theoretical argument for the safety of these experiments: *if* we really managed to make small black holes, then they would evaporate almost immediately because of a process called Hawking radiation. The lighter a black hole is, the warmer it is, and the faster it evaporates. In a particle collision experiment, this evaporation would leave a spectacular fingerprint. It's a lot like a steaming cup of tea on the breakfast table, which gives off warmth (photons), except that when a mini black hole evaporates, it has a very democratic policy of giving off all the particles in the Standard Model. This exotic blend of particles has a unique signature, most comparable to fireworks going off on New Year's Eve. It would be a spectacular sight, not to be missed.

The collisions at the LHC have not yet produced any evidence that extra dimensions exist. We haven't seen any mysterious disappearances or mini black holes. But we hope that in a few years we can generate enough energy to lift the glass off the surface of the table. The experimental facts familiar to us now suggest that the extra dimensions have a size or structure that remains invisible to us at the LHC energy level. Meanwhile, the strange weakness of gravity remains an unsolved problem, and the idea that extra dimensions could be the explanation is still as fascinating as ever.

Aren't there any alternative ideas about gravity? Sure, but not so many. A few years ago, one of the best known Dutch theoretical physicists, Erik Verlinde of the University of Amsterdam, presented the idea that gravity is not a fundamental force of nature at all, but something more like an emergent "force" resulting from an underlying ordering principle in space. When milk mixes with coffee, there's no force of nature that says they have to mix evenly—it just happens spontaneously. This is an example of the statistical phenomenon known as entropy, which tells us that a fairly uniform mixture is a much more likely result than a separate cloud of milk floating

around in the black coffee. The same principle could underlie the attraction between objects with mass. Each bit of space
may contain a certain amount of information trying to orga-
nize itself. Matter that's close together is apparently better organized, so it looks as if there's actually something bringing
the matter together.

If this is true, it would give us a whole new perspective
on space-time and the peculiar role that gravity currently
plays in physics. And not only that: according to Verlinde's
proposal, the dark matter we see as one of the biggest problems in present-day physics and astronomy is no more than an
illusion. It's an intriguing and very thought-provoking theory,
but we can't yet say whether it's right or not. We don't know
the nature of the information or the mechanism that organizes
it, and everyone's still searching for an experimental prediction that will enable us to test this new idea. Exciting times!
And if we find out that dark matter is an illusion, and that the
theory of relativity and our understanding of space-time are
based on a deeper mechanism, then the illustrious list of gravity greats—Isaac Newton, Albert Einstein—may soon include
a new name: Erik Verlinde.

Amazing new ideas like this one are the fuel of scientific
progress, and in the end, it's experimentation that must determine which theory survives.

THE REALLY BIG QUESTIONS ABOUT SPACE AND TIME

There are questions and there are questions. The "deep," fundamental questions go to the heart of our place in the universe, and thinking about them can make us uneasy. If we ask
ourselves what the really big questions are in physics, here are
the ones that come to mind for me. They're questions I don't

really believe we ever *can* answer, but all the same, they're the ones that keep me awake at night.

What is space made of, and how much is there? How large is the universe? We know that the universe keeps expanding faster and faster, even though we don't even know how to make a single cubic meter of extra space. In fact, we don't even know what space itself is made of. And I myself realized only recently that we have no idea how large the universe truly is. We know that everything we can see today seems to have come from a single point, but we also know that there's a lot more universe beyond what we can see now, and that we never will be able to see all that space. The volume of the universe could even be infinite. It surprises me how casually my fellow scientists sometimes talk about this: "Finite or infinite—both are possible and we'll never know, so quit your worrying." But I do worry. I can work with infinities in mathematics—fine. But I just can't imagine that I, as a human being, live in an infinitely large universe. That just can't be true. Still, whatever the case may be, there's an even bigger question: how did space come into existence in the first place?

Does (space-)time have a beginning? This is another example of the type of question that somehow you can't seriously ask in the scientific world. Within our present concept of the evolution of the universe, the question is almost nonsensical, because by definition there was nothing before the Big Bang, the same way there's nothing to the north of the North Pole. But that answer strikes me as a little too easy. Like trying to make a quick getaway from wonder and astonishment by burying yourself in math. Think how bizarre it is that such things as time and space exist at all. Right? Of course it's weird to think that time once had a beginning. Because what defines that beginning? But the alternative—that time never began, or that it's cyclical—is just as ridiculous from our human perspective. The beginning, the whole essence of time, is a riddle that humanity just can't get its head around. Yet the question keeps gnawing at us, and the longer I think about it, the more uneasy I feel.

The Adventure Continues

The curiosity of human beings is still nowhere near satisfied. We have seen a small number of the many questions in particle physics (including astroparticle physics), but every scientific field has its fascinating puzzles and unexplored areas. Although the real scientific breakthroughs take place only rarely, they're clearly what motivates all of us. And in the last analysis, those leaps of scientific progress arise from the wonder and astonishment of people who search for answers, systematically gathering data and puzzle pieces.

If we've learned anything over the past hundred years, it's that nature keeps surprising us time after time. So there are still plenty of dreams for us to dream. But eventually, we'll have to step over the edge, into the new world. That's the only way—through experimentation—that we can discover how nature works, see which theoretical ideas are false or true, and then work out all the implications for the way we see the world.

People are still looking up at the skies, just as they did at the dawn of humanity, and asking the same questions. Where do we come from? And how does it all work? We have no idea how much closer we've come to the ultimate answer, but the adventures that brought us this far were well worth the effort. I hope that, like me, you're looking forward to exploring the unknown territories ahead of us. As you've seen, many scientists, including me, are hopeful that in the years ahead we'll be able to answer some of the big questions. We have unsolved mysteries, we have beautiful dreams and visions of the new world, and we're bursting with ideas for experiments that may really take us there. Let's step over the edge together as soon as we can and discover what new particles, forces, and concepts might show up along the way—if we succeed in taking that step, of course. Off we go!

Acknowledgments

Putting all the peculiarities of the world of elementary particles down on paper is no easy task. Especially not when it all has to happen after work and be both scientifically sound and accessible to the interested layperson. The fact that the book, despite these hurdles, has become a reality and is now in your hands is entirely thanks to the support and perseverance of my editor, Bertram Mourits. Thank you!

I'd also like to thank a few other people: Ernst-Jan Buis and Niels Tuning, for acting as sounding boards for my ideas, and Jordy de Vries, for checking "the big formula." Rody van Vulpen was my reader every time I finished penning another chapter, and thanks to Sarina van der Ploeg, we made lots of improvements to the text in the final stage. I'm delighted that Serena Oggero agreed to do the illustrations, and Kees Huyser fit them neatly into the book.

I would like to thank Joe Calamia of Yale University Press for believing in this book and deciding to publish this English edition. Also a big thank you to David McKay, who translated the book into English, making it possible for many more people to read about the wonders of the world of elementary particles.

Obviously, nothing gets finished without support from the home front. Julia, Cleo, and Olivia, I'm so happy to have you in my life. And ladies, I'm really done now.

Index

Page numbers followed by "f" or "t" denote figures and tables, respectively.